T0311799

OXFORD CLASSIC TEXTS IN
THE PHYSICAL SCIENCES

THE PHYSICS OF RUBBER ELASTICITY

Third Edition

BY

L. R. G. TRELOAR

CLARENDON PRESS · OXFORD

This book has been printed digitally and produced in a standard specification in order to ensure its continuing availability

OXFORD
UNIVERSITY PRESS

Great Clarendon Street, Oxford OX2 6DP

Oxford University Press is a department of the University of Oxford.
It furthers the University's objective of excellence in research, scholarship,
and education by publishing worldwide in

Oxford New York

Auckland Cape Town Dar es Salaam Hong Kong Karachi
Kuala Lumpur Madrid Melbourne Mexico City Nairobi
New Delhi Shanghai Taipei Toronto
With offices in
Argentina Austria Brazil Chile Czech Republic France Greece
Guatemala Hungary Italy Japan South Korea Poland Portugal
Singapore Switzerland Thailand Turkey Ukraine Vietnam

Oxford is a registered trade mark of Oxford University Press
in the UK and in certain other countries

Published in the United States
by Oxford University Press Inc., New York

ISBN 978-0-19-857027-1

PREFACE TO THE THIRD EDITION

THE preparation of the Third Edition of this book has presented problems which were not encountered either with the First or with the Second Edition. The expansion of the subject during the last 16 years has involved a problem in the selection and arrangement of material to which there is no completely satisfactory solution. As a guiding principle I have assumed that the primary objective should be to provide a logical and reasonably detailed presentation of the main developments in the field of the *equilibrium* elastic properties of rubber (including the photoelastic and swelling properties), together with the associated theoretical background. In consequence it has been necessary to eliminate the last two chapters of the Second Edition, dealing respectively with stress-relaxation and flow and dynamic properties. The two chapters relating to crystallization have also been removed, though some references to this subject have been included in an enlarged Chapter 1. Despite its great inherent interest, particularly in relation to the historical development of the physics of rubber elasticity, the subject of crystallization in rubber is now seen to be incidental rather than fundamental to the main theme of this book, and its proper treatment would require an extensive discussion of crystallization in polymers other than rubber. A number of authoritative treatments of this wider subject are already in existence.

The main advances in more recent years have been in the thermodynamic analysis of rubber elasticity and in the essentially separate development of the phenomenological (i.e. non-molecular) approach to the subject. To take account of the former it has regrettably been necessary to divide the treatment of the thermodynamics into two parts, the first (elementary) being contained in Chapter 2, and the second (advanced) in the final chapter. The previous Chapter 8 (on phenomenological theory) has been expanded into three separate chapters (Chapters 10, 11, and 12), of which Chapter 11 contains essentially new material.

Inevitably these changes will to some extent reduce the attractiveness of the book for the student who wishes to acquire a broad knowledge of the whole range of physical phenomena associated with rubber. It can only be hoped that this loss will be more than offset by the greater value of the book as a critical review of the subject of rubber elasticity in the more restricted sense, in which field no comparable work is readily available.

L.R.G.T.

Department of Polymer and Fibre Science,
University of Manchester Institute of Science and Technology

PREFACE TO THE FIRST EDITION

IT IS sometimes considered unnecessary for those engaged in the practical development of industrial processes to concern themselves with the so-called theoretical aspects of their subject. On examination, it is usually found that exponents of this point of view are not entirely consistent, for in any type of work involving experimentation it is impossible to get along without some sort of theory, however limited or *ad hoc* it may be. My excuse for doing the work which I do (of which this book is one aspect) is that I always believe that if one is going to have a theory at all one may as well take some trouble to find the one which most nearly represents the known facts.

In the subject of rubber elasticity it is not easy to discover from the mass of literature, often of a rather mathematical character, what are the generally accepted theories. In this book I have therefore attempted to convey (in not too mathematical language) the fundamental concepts of the subject, and to present the whole in a more or less consistent form. In this task I have admittedly given expression to my own point of view, and I have drawn freely on the work of my associates at the British Rubber Producers' Research Association. I cannot hope to acknowledge the many who have helped me by the discussion of particular sections, but I should like to mention particularly Dr. G. Gee, Director of the B.R.P.R.A., who read and criticized the manuscript in detail, my colleague Mr. R. S. Rivlin, who gave me the benefit of his unpublished ideas and works, and Dr. K. Weissenberg, with whom I was able to discuss the final chapter.

I should also like to thank the Board of the B.R.P.R.A. for encouraging me to undertake this work, and for the provision of facilities for its execution.

L.R.G.T.

British Rubber Producers' Research Association,
Welwyn Garden City

ACKNOWLEDGEMENTS

THE National Bureau of Standards (Washington) with N. Bekkedahl for Figs 1.3 and 1.6 and with L. A. Wood for Fig. 1.7; The American Chemical Society for Figs 2.2, 2.3, 2.4, 2.7, 2.8 and 2.11; The American Chemical Society with P. J. Flory for Figs 7.5 and 8.1 and with M. Shen for Figs 13.5 and 13.6; The Soviet Academy of Sciences with A. P. Aleksandrov for Fig. 1.4; The Institution of the Rubber Industry with L. Mullins for Fig. 1.5 and with G. Gee for Fig. 7.8; Applied Scientific Research (Martinus Nijhoff) for Fig. 1.9; Helvetica Chimica Acta for Fig. 2.1; The Chemical Society for Figs 1.10, 5.4, 5.5, 5.6, 5.7, 5.8, 5.9, 7.1, 7.10, 7.11, 7.12, 9.2, 9.3, 9.6, 9.6, 9.7 and 9.8; The Chemical Society with L. Mullins for Figs 5.11, 5.12, and 5.13, with G. Gee for Figs 7.2, 7.3, and 7.6, with D. W. Saunders for Figs 9.11, 9.12, 9.13, 9.14, and 9.15, and with G. Allen for Figs 13.1, 13.2, 13.3, and 13.10; The American Institute of Physics with L. A. Wood for Fig. 2.9, with E. Guth for Fig. 2.10 with R. F. Landel for Fig. 11.2, and with M. Shen for Fig. 13.4; John Wiley and Sons Ltd., with J. Scanlan for Fig. 4.5, with L. Mullins for Figs 6.13, 6.14, 7.7, 8.3, 8.5, and 8.6; with M. C. Morris for Fig. 6.14, with W. F. Watson for Fig. 8.4, with A. N. Gent for Fig. 9.16, with D. W. Saunders for Fig. 9.18, with S. Kawabata for Figs 11.3, 11.4, 11.8, and 11.9, and with R. G. Christensen for Fig. 13.9, I.P.C. Business Press Ltd. for Figs 7.13, and 13.8, and with G. Allen for Fig. 13.7; The Royal Society with D. W. Saunders for Figs 10.1, 10.8, 12.9, and 12.10, and with R. W. Ogden for Fig. 11.1; The Institute of Physics for Figs 9.9, 9.10, 10.5, 10.6, 10.7, 11.5, 11.6, 11.7, and 11.11; Marcel Dekker Inc. with S. Kawabata for Fig. 11.10.

CONTENTS

1

GENERAL PHYSICAL PROPERTIES
OF RUBBERS

1.1. What is a rubber?

THE original material of commerce known as rubber (or more precisely 'india-rubber') is obtained in the form of latex from the tree *Hevea Braziliensis*. The more expressive term 'caoutchouc', derived from the Maya Indian words meaning 'weeping wood', in reference to the exudation of the latex from a wound in the bark (Le Bras 1965), has been retained by the French, and transliterated into other European languages. The word *rubber* is derived from the ability of this material to remove marks from paper, to which attention was drawn by the chemist Priestley in 1770 (Memmler 1934, p. 3). In current usage the term rubber is not restricted to the original natural rubber, but is applied indiscriminately to any material having mechanical properties substantially similar to those of natural rubber, regardless of its chemical constitution. The more modern term *elastomer* is sometimes employed in relation to synthetic materials having rubber-like properties, particularly when these are treated as a sub-class of a wider chemical group. However, in the present work the more popular usage will be followed. It will generally be obvious from the context whether the word rubber is used in the general or in the more restricted sense; in cases where confusion might arise it will be sufficient to refer to *natural* or *Hevea* rubber.

The reasons for this choice are not entirely verbal. It is at least equally justifiable from the scientific standpoint to define a rubber in terms of its physical properties as in terms of its chemical constitution. Indeed, in the present work, we shall be concerned very much more with those fundamental structural aspects in which all rubbers may be considered to be essentially the same than with the more detailed specific features in which they differ from one another. The emphasis will be placed mainly on *rubber-like elasticity* as a phenomenon associated with the *rubber-like state* of matter.

The most obvious and also the most important physical characteristic of the rubber-like state is of course the high degree of
deformability exhibited under the action of comparatively small
stresses. A typical force–extension curve for natural rubber is
shown in Fig. 1.1; the maximum extensibility normally falls within

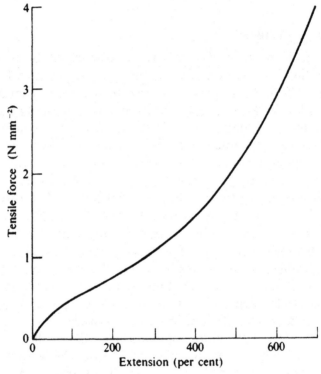

FIG. 1.1. Typical force–extension curve for vulcanized rubber.

the range 500–1000 per cent. The curve is markedly non-linear (i.e.
Hooke's law does not apply), hence it is not possible to assign a
definite value to Young's modulus except in the region of small
strains. In this region its value (represented by the tangent to the
curve at the origin) is of the order of $1 \cdot 0 \, \text{N mm}^{-2}$. These
properties—high extensibility and low modulus—are to be con-
trasted with the properties of a typical hard solid (e.g. steel), for
which the value of Young's modulus is $2 \cdot 0 \times 10^5 \, \text{N mm}^{-2}$ and the
corresponding maximum elastic (i.e. reversible) extensibility about
$1 \cdot 0$ per cent or less. There is thus an enormous difference between

rubbers on the one hand and ordinary hard solids (crystals, glasses, metals, etc.) on the other.

Thermoelastic effects

In addition to these familiar mechanical properties rubber also possesses a number of other less well-known properties, namely, the thermal or thermoelastic properties, which are of even greater scientific significance. The study of these properties dates from the beginning of the last century, when Gough (1805) made the following two observations, i.e.

(1) that rubber held in the stretched state, under a constant load, contracts (reversibly) on heating; and

(2) that rubber gives out heat (reversibly) when stretched.

Gough's conclusions were confirmed some 50 years later by Joule (1859), who worked with the more perfectly reversible vulcanized rubber which had become available since the time when Gough's original experiments were carried out. The two effects referred to are usually known as the Gough–Joule effects. An example of the second, taken from Joule's publications, is reproduced in Fig. 2.10 (p. 38); this shows the rise of temperature due to the evolution of heat on stretching up to an extension of 100 per cent.

These thermoelastic effects are not peculiar to natural rubber, but are characteristic of the rubber-like state, being observed in a wide variety of synthetic rubbery polymers.

1.2. Chemical constitution of rubbers

Natural rubber is essentially a hydrocarbon, whose constitution was established by Faraday (1826) to be consistent with the formula $(C_5H_8)_n$. The rubber exists in the latex in the form of small globules, having diameters in the range $0 \cdot 1 - 1 \cdot 0$ μm, suspended in a watery liquid or serum, the concentration of the rubber being about 35 per cent. The rubber particles would coalesce, of course, were it not for a layer or sheath of non-rubber constituents, principally proteins, which is adsorbed on their surfaces and functions as a protective colloid. From this latex the solid rubber may be obtained either by drying off the water or by precipitation with acid. The latter treatment yields the purer rubber, since it leaves most of the non-rubber constituents in the serum.

Chemically, the rubber hydrocarbon is a polymer of isoprene (C_5H_8) built up in the form of a continuous chain having the

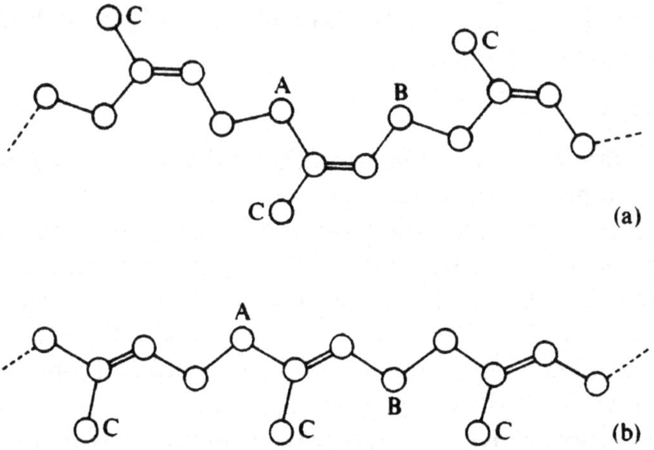

FIG. 1.2. The structure of the molecule of (a) *Hevea* rubber and (b) gutta-percha. A–B = isoprene unit. C = methyl group.

structure shown in Fig. 1.2. The succession of isoprene units in the chain is perfectly regular, with every fourth carbon atom in the chain carrying the methyl (CH₃) side-group. The presence of the double bond is very significant, since it is this that largely determines the chemical reactivity of the molecule and its ability to react with sulphur or other reagents in the vulcanization process. The double bond is also responsible for the susceptibility of the rubber molecule to oxidation or other degradative reactions leading to a deterioration of physical properties (aging).

The structure of gutta-percha, the other natural polymer of isoprene, differs slightly but significantly from that of rubber. As will be seen from Fig. 1.2, the difference lies solely in the arrangement of the single C—C bonds with respect to the double bonds in the chain backbone. In rubber the single bonds lie on the *same* side of the double bond, forming the so-called *cis*-configuration, whilst in gutta-percha they lie on opposite sides of the double bond, giving the *trans*-configuration. One consequence of this difference is that gutta-percha crystallizes more readily than rubber; it is in fact crystalline at room temperature, becoming rubber-like only when heated above the crystal melting point, namely, 65 °C.

Although the two single bonds adjacent to the double bond remain permanently fixed in a single plane (whether in the *cis*- or *trans*-configuration), the remaining single bonds are not thus fixed but are subject to rotation out of the plane formed by neighbouring bonds, as will be discussed in detail later. The structural forms

TABLE 1.1
Structural formulae of some typical rubbers and related materials

$-CH_2-\underset{\underset{CH_3}{|}}{C}=CH-CH_2-$ Polyisoprene (natural rubber, gutta-percha)

$-CH_2-CH=CH-CH_2-$ Polybutadiene

$-CH_2-\underset{\underset{Cl}{|}}{C}=CH-CH_2-$ Polychloroprene (Neoprene)

$-CH_2-\underset{\underset{Cl}{|}}{CH}-$ Polyvinyl chloride

$-CH_2-CH-$ Polystyrene

$-CH_2-\underset{\underset{CH_3}{|}}{\overset{\overset{CH_3}{|}}{C}}-$ Polyisobutylene (basis of 'butyl' rubber)

$-CH_2-CH=CH-CH_2 \! \mid \! CH_2-CH-$ †Butadiene-styrene (BSR) rubber

$-CH_2-CH=CH-CH_2 \! \mid \! CH_2-\underset{\underset{CN}{|}}{CH}-$ †Butadiene-acrylonitrile ('nitrile') rubber

$-O-\underset{\underset{CH_3}{|}}{\overset{\overset{CH_3}{|}}{Si}}-$ Polydimethyl siloxane (silicone rubber)

$-CH_2-$ Polyethylene (polythene)

$-CH_2-\underset{\underset{CH_3}{|}}{CH}-$ Polypropylene

$-CH_2-\underset{\underset{COOCH_3}{|}}{\overset{\overset{CH_3}{|}}{C}}-$ Polymethyl methacrylate (Perspex)

† In these copolymers the respective monomer units occur in a random sequence along the chain.

depicted in Fig. 1.2 are thus to a certain extent schematic; in practice the molecule does not have the linear planar conformation which these diagrams suggest.

Synthetic rubbers

Examples of some of the more important synthetic rubbers are given in Table 1.1. The most widely used synthetic general-purpose rubber is a butadiene–styrene copolymer (SBR), developed on a large scale in U.S.A. during World War II to make good the loss to the enemy of the major rubber-growing areas of the world (Malaya and the Dutch East Indies). Butyl rubber, based on polyisobutylene, which has a very low gas permeability, has important applications for inner tubes, etc. Polychloroprene (Neoprene) has superior resistance to degradation, compared with natural rubber, with the further advantages of lower oil-absorption and lower flammability; these properties render it attractive for engineering applications, particularly where oil or petrol contamination is likely. Another oil-resistant rubber is 'nitrile' rubber—a copolymer of butadiene and acrylonitrile; this material, however, has poor resilience at low temperatures (cf. Fig. 1.5, p. 15). Polybutadiene itself is highly resilient (i.e. has very low hysteresis losses) over a wide temperature range, as are also the silicone rubbers, based on the Si—O chain structure; the latter are also particularly resistant to chemical degradation, and may be used over the range $-55\,°C$ to $+300\,°C$ (Houwink 1949). Another material of interest, e.g. in tyres, is the so-called ethylene–propylene terpolymer; this is a copolymer consisting essentially of ethylene and propylene, but with a small proportion of a third ingredient.

1.3. Early theories of rubber elasticity

Early attempts to account for the mechanical properties of rubber in terms of classical concepts of the molecular structure of matter encountered overwhelming difficulties. In the first place natural rubber (the only type of rubber then available) appeared both unique and enigmatic. Yet its molecular constitution as then understood could hardly have been simpler. Other known hydrocarbons of apparently entirely similar chemical constitution were invariably ordinary liquids or solids. What was it about the rubber hydrocarbon that should render it so different from, say, terpentine, which has an identical empirical formula?

Quite apart from this difficulty, the classical solid was envisaged as an array of atoms (or molecules) maintained in fixed mean relative positions by well-defined interatomic forces. Application of a stress (e.g. a tensile force) to such a structure leads to a disturbance of the equilibrium which calls into play a counterbalancing internal stress. Owing to the very strong dependence of such interatomic forces on the distance between atoms, such a mechanism is incapable, even theoretically, of supporting extensions of magnitude greater than about 10 per cent; it cannot conceivably serve as a basis for the interpretation of deformations of 100 times this magnitude.

The basic difficulty can in principle be circumvented, without entirely breaking away from classical concepts, by postulating either some sort of open network structure, or alternatively a helical or coil-spring type of molecule. In either system large total deformations may be obtained without the introduction of large strains into the elastic elements of the structure. Examples of the first type of theory were the 'two-phase' theories, among which that of Ostwald (1926) was perhaps the most plausible. This attributed the essential elasticity to a sort of network of micelles or molecular aggregates based on proteins or resins derived from the outer sheath of the latex globule, this network being suspended in a semi-liquid medium formed of rubber hydrocarbon of lower molecular weight. Among theories of the second type, the helical-spring theory of Fikentscher and Mark (1930) was the most notable. In this, the retraction tendency was associated with the residual forces between neighbouring turns of the helix which was believed to represent the configuration of the polyisoprene chain.

A rather similar theory by Mack (1934) envisaged a folded configuration of the molecule which was maintained by the agency of forces between neighbouring hydrogen atoms. This model permitted an extension of 300 per cent in passing from the closely packed to the fully extended form by rotation about single bonds in the chain structure. A further extension of the assembly of molecules, leading to a total extensibility of 600 per cent, was accommodated by allowing for a rotation of the extended chains from their initially random directions into the direction of the applied strain.

1.4. The kinetic theory of elasticity

The theories referred to above, and others of a similar character, sought to explain the long-range elasticity in terms of the only available rubber in common use at the time, i.e. natural rubber. Their interpretation of the mechanical properties of this material was unconvincing, and furthermore, they were completely incapable of accounting for its even more remarkable thermoelastic properties. It is significant that the first real progress towards a proper understanding of rubber elasticity owed much to the recognition that natural rubber was no longer to be regarded as in a class by itself but that it was very closely related in both structure and properties to a number of other materials, e.g. gelatin, muscle fibres, silk, etc., broadly classed as colloids, but widely diverse in chemical constitution. Thus, for example, Wöhlisch (1926) drew attention to the similarity between the contraction of tendons (collagen) on heating and the contraction of stretched (crystalline) raw rubber, which he sought to explain in terms of the thermal agitation of crystals, micelles, or rod-like molecules. Later, with the refinement of methods of measuring very high molecular weights and the emergence of the concept of a *polymer* as a genuine chemical entity whose molecular weight could be in the range 100 000–1 000 000 (a typical value for rubber being 350 000), the full significance of the form and dimensions of the molecule began to be appreciated. It gradually became evident that a molecule of such dimensions could no longer be regarded as a geometrically rigid structure held in a fixed configuration by static internal forces, but that some degree of flexibility, associated with internal vibrations and rotations having their origin in thermal fluctuations, was to be expected. Thus, for example, Haller (1931) calculated that the thermal fluctuation of bond lengths and valence angles in a paraffin chain could lead to a considerable curvature of the chain axis. However, this calculation ignored the much more important consideration of rotation about bonds.

The first exponents of the now generally accepted theory of rubber elasticity were Meyer, von Susich, and Valko (1932) and Karrer (1933), though the latter author was thinking primarily of muscle. Meyer *et al.* based their theory on the consideration that the thermal energy of the atoms of a long-chain molecule will lead to greater amplitudes of vibration in the direction perpendicular to the chain than in the direction of the chain itself, since the lateral forces between chains are much weaker than the primary valence forces within the chain. The effect of this dissimilarity would be to produce a repulsive pressure between parallel or extended chains, which would have the effect of tending to draw the ends of the chains together and could thus be regarded as equivalent to a longitudinal tension. As a result a stretched rubber should retract 'until an irregular statistically-determined arrangement of the molecules and their parts is brought about, in which condition molecular impacts no longer have a resultant directional effect'. The quotation shows that Meyer *et al.* clearly appreciated the relation between rubber-like elasticity and the ability of the chain to take up an irregular or statistically determined form by virtue of the interchange of energy between its constituent elements and the surrounding atoms. This is the fundamental concept of the now generally accepted theory of rubber elasticity.

The authors showed that their kinetic theory led directly to the conclusion that the tension in stretched rubber (at constant length) should be proportional to the absolute temperature—a result which was confirmed in later experiments by Meyer and Ferri (Fig. 2.1, p. 25).

Though using as an argument in favour of their theory the similarity in elastic properties exhibited by various materials (gelatin, cellulose, silk, etc.), Meyer and his associates recognized the quantitative differences between particular materials and drew attention to the significance of crystallization as a factor determining the range of extensibility and other mechanical properties in those polymers in which it was believed to occur.

Karrer (1933) attempted to explain the properties of muscle fibres on similar lines and, like Meyer, regarded muscle as a member of a class of long-chain structures which included rubber. He noted that a long-chain molecule in which internal rotation about bonds can take place will be subject to the Brownian motion of the various elements of its structure and will consequently take up a variously contorted shape unless constrained by a tensile force applied to its ends. This tendency of a chain to take up a random conformation he related to the principle of 'maximum mechanical chaos', which is equivalent to the concept of maximum entropy developed by Meyer. In the normal or resting state the protein molecules in the muscle fibre were assumed to be held in a parallel extended alignment by chemical forces; the process of retraction or activation of the muscle was then assumed to be initiated by chemical changes leading to the loosening of these intermolecular forces.

By thermodynamic reasoning Karrer showed that the performance of work by the retracting muscle should be accompanied by an absorption of heat, or, in an adiabatic process, by a fall of temperature. The calculated amount of this fall for the muscle in a frog's leg was 0·006 °C.

This interest in the thermodynamic implications of the phenomenon of rubber-like elasticity may be regarded as a return to the point reached in the pioneer investigations of Joule in the mid-nineteenth century. As has already been noted Joule made measurements of the heat changes developed in the extension of rubber, and Kelvin (1857) demonstrated the connection between these thermal effects and the effect of temperature on the elastic retractive force discovered by Gough (1805). It is stated by Meyer (1939), and also by Guth, James, and Mark (1946), that Joule recognized that the contraction of stretched rubber is due to the motion of its constituent particles. This may have been so, but so far as the author has been able to discover, it is nowhere explicitly stated in his writings.† In any case, Joule was not in a position to develop any kind of more detailed interpretation of the phenomenon, owing to the totally inadequate knowledge of the chemical structure of rubber at that time.

Even after the publication of Meyer's theory, many years elapsed before scientists generally were prepared to acknowledge its superiority over rival interpretations of rubber elasticity. This reluctance to accept what is now generally regarded as a major scientific advance may be attributed partly to a very proper scepticism with respect to a subject which had become overburdened with somewhat fanciful theories with little experimental backing, partly to the difficulty of separating the essential phenomena of elasticity from the subsidiary, though quantitatively very significant, thermomechanical effects associated with strain-induced crystallization, and partly to the revolutionary nature of the kinetic theory itself, this last factor being closely related to the corresponding reluctance to accept the concept of a high polymer in any form whatever. However, with the advent of a wide range of synthetic rubbers, the significance of the theory became rapidly more apparent, and the concept of rubber-like elasticity as an attribute of a certain general

---

† In private correspondence Meyer writes, 'Joule must have drawn the conclusion himself, but he does not say it clearly'.

type of molecular structure rather than of the specific constitution of the polyisoprene chain became more widely accepted.

1.5. Cross-linking and vulcanization: network theory

General conditions for rubber-like elasticity

Implicit in the preceding discussion is the implied assumption that the inherent elasticity of the long-chain molecule is sufficient in itself to confer highly elastic properties on a material in bulk. This, however, is only one, albeit the most fundamental, of the necessary conditions. A more realistic consideration of this problem must take into account not only the properties of the molecule in isolation but also the way in which the individual molecules are held together so as to form a coherent structure. Taking these considerations into account, we are led to the conclusion that for a material to exhibit rubber-like properties the following three requirements must be satisfied:

(1) the presence of long-chain molecules, with freely rotating links;

(2) weak secondary forces between the molecules;

(3) an interlocking of the molecules at a few places along their length to form a three-dimensional network.

The first of these conditions has already been dealt with. The second arises from the consideration that, if the individual chain is to have the freedom to take up the variety of statistical conformations upon which the phenomenon of rubber-like elasticity ultimately depends, its motion must not be impeded by the surrounding molecules; this implies that the forces between the molecules (secondary forces) shall be weak, as in a liquid. Indeed, apart from the fact that successive segments are permanently connected to each other by primary chemical bonds, the intermolecular forces in a rubber are in no way different from those existing in a typical liquid. However, if this were the *only* requirement, the material would in fact behave as a liquid, and not as a solid. The third of the above conditions is introduced in order to overcome this difficulty. By the introduction of a certain number of cross-linkages or junction points between the chains at a very few points along their length it is possible to produce a coherent network in which all the molecules are linked together and hence can no longer move independently as in a liquid. Owing to the great length of the chains, the number of such points of cross-linkage required to achieve this result (theoretically two per chain) is not sufficient to interfere significantly with the requisite local freedom of movement or statistical fluctuations of the individual chains.

Network formation

The necessary cross-linkages between chains are normally introduced by the process of vulcanization, which is a chemical reaction with sulphur, originally discovered by Charles Goodyear in 1839 (Memmler 1934). Though other reagents are frequently used in current industrial practice, vulcanization has remained an essential requirement of rubber technology from that time until the present day. Rubber articles are extruded or moulded into their required form at a suitably elevated temperature while the rubber is in a semi-liquid or plastic condition; the final form is then fixed, and the required elasticity or rigidity secured, by cross-linking of the chains.

In dealing with the subject of rubber elasticity we shall be concerned in general not with the original raw rubber but with the vulcanized material, which is more perfectly elastic (i.e. reversible) in its behaviour than raw rubber. In passing, however, it should be mentioned that in raw rubber an effect similar to that of chemical cross-linking is obtained from the complex geometrical entanglements between chains, which produce a local enhancement of the residual (van der Waals) forces. Under prolonged loading such 'entanglement-cohesions' or 'physical' cross-linkages will slowly break down, giving rise to the phenomena of stress-relaxation and creep, but for short times of stressing a degree of elasticity not very different from that of vulcanized rubber may be displayed by the unvulcanized material.

One of the most important developments in the statistical theory has been its application to the problem of the network of long-chain molecules as it exists in a vulcanized rubber. This application, which is fully dealt with in Chapter 4, has led to the derivation of specific stress–strain relations for various types of strain, which may be compared with experimental observations. This development has been particularly fruitful, and has transformed our outlook on the whole subject of the structure and physical properties of rubbers. Its influence has extended far beyond the consideration of the purely mechanical properties. For example, by a slight extension of its basic concepts it has been found possible to adapt the theory so as to provide a quantitative interpretation of such additional phenomena as the swelling of a rubber in organic liquids, and the development of double refraction under stress (photoelasticity).

1.6. The glass–rubber transition

The rubber-like state, as we have seen, depends on the possibility of random thermal motion of chain elements by rotation about single bonds in the chain backbone. In any real material rotation cannot be completely free from restrictions imposed by the presence of neighbouring groups of atoms either in the same molecule or in neighbouring molecules. The degree of freedom of rotation will be a function of the relative values of the thermal energy of the rotating group and the potential barrier that has to be overcome in order that rotation may occur. The probability that a given group will have sufficient energy to enable it to surmount a potential barrier ε will be governed by a Boltzmann factor of the type $\exp(-\varepsilon/kT)$, and will therefore increase rapidly with increase in temperature. Conversely, on lowering the temperature a point will be reached at which rotation will no longer take place at an appreciable rate. In this state the material ceases to behave like a rubber and becomes hard and rigid like a glass.

The transition from the rubber-like to the glassy state is a phenomenon which is encountered in all rubbers, whether vulcanized or unvulcanized, though the temperature at which this transition occurs naturally depends on the chemical constitution of the molecule. The transition temperature is about $-71\,°C$ for unvulcanized natural rubber, and a few degrees higher for the vulcanized material. The geometrical structure is not affected by the transformation, being still of the random or amorphous type; this is shown by the X-ray diffraction pattern, which has the form characteristic of a liquid or glassy structure, namely a broad diffuse ring or 'halo'.

The transition to the glassy state is accompanied by changes in certain other physical properties in addition to the changes in elastic properties. Of these the most important is the change in expansion coefficient. This is shown in the top curve in Fig. 1.3, representing the volume–temperature relationship for raw rubber in the amorphous state. The change in slope at the transition temperature corresponds to a considerable increase in expansivity which is usually interpreted as a direct result of the increase in molecular

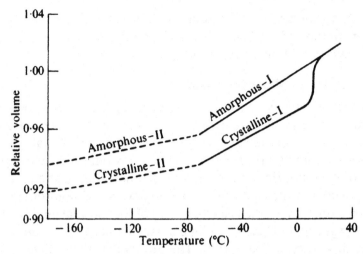

FIG. 1.3. Volume–temperature relationship for purified rubber, showing second-order transition at −72 °C and crystal melting at 11 °C. (Bekkedahl 1934.)

mobility associated with the rubber-like or quasi-liquid structure. Since, however, the glass–rubber transition is not a *structural* change there is no *first-order* change in specific volume (or density) such as occurs, for example, in the transformation from the amorphous to the crystalline state (see below). For this reason the glass–rubber transition is sometimes referred to as a *second-order* transition. Another property which undergoes a similar type of change at the same temperature is the heat content or enthalpy; this is revealed by an increase in the specific heat, and is similarly indicative of an increase in molecular mobility. ·

Gradualness of transition

The change in mechanical properties from the glassy to the rubbery state as the temperature is raised is not as sudden as the change in expansivity shown in Fig. 1.3, but takes place over a range of temperature of 50 °C or more. This is shown, for example, in Fig. 1.4, which represents the variation of amplitude of deformation, under an oscillatory stress of constant amplitude, with increase of temperature. Below −70 °C the deformation is very small, corresponding to a modulus of elasticity of the order 10^9 N m^{-2}. As the temperature rises it increases at first rapidly, and then more slowly as the full rubbery deformation, corresponding to a modulus of the order 10^6 N m^{-2}, is approached. In the intermediate transition region the elastic properties are very poor, owing to the high

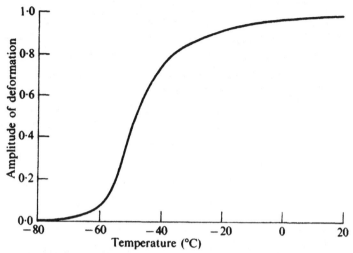

FIG. 1.4. Amplitude of deformation as function of temperature. Frequency $\frac{1}{60}$ Hz. (Aleksandrov and Lazurkin 1939.)

internal viscosity (hysteresis) resulting from the limited mobility of chain segments.

Rebound resilience

The phenomena encountered in the transition region are directly reflected in the property of rebound resilience. If there were no losses the rebound of a ball would be 100 per cent. For natural

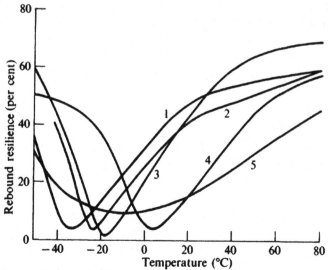

FIG. 1.5. Rebound resilience as function of temperature. (Mullins 1947.) 1. Natural rubber. 2. SBR (butadiene-styrene). 3. Neoprene (poly-chloroprene). 4. Hycar OR 15 (butadiene-acrylonitrile). 5. Butyl rubber.

rubber at room temperature the rebound resilience is around 60 per cent, corresponding to the dissipation of some 40 per cent of the elastic strain energy by internal viscosity. As the temperature is lowered the resilience falls, reaching a very low minimum at about −35 °C, after which it again rises steeply as the glass-hard state is approached. (Fig. 1.5). The temperature of minimum resilience (maximum energy loss) corresponds to the middle of the transition region shown in Fig. 1.4, where the properties are changing most rapidly; this is some 30 °C above the glass transition temperature as determined from expansivity measurements.

Glassy polymers

The temperature at which the rubber–glass transition occurs is an extremely important property of a polymer. For general-purpose rubbers it is important that this temperature shall be sufficiently low to avoid undesirable energy losses over the range of temperature likely to be encountered in actual use. The factors which determine the transition temperature are primarily the strength of the inter-molecular forces and the degree of flexibility of the chain. Bulky or highly polar side-groups introduce steric hindrances to internal rotation about bonds as well as increasing the intermolecular forces; such groups therefore tend to raise the glass transition temperature. These effects are illustrated by the two most common of the glassy polymers, namely, poly(methyl methacrylate) (Perspex) and poly-styrene, whose structures are given in Table 1.1 (p. 5). Each of these has a glass transition temperature in the neighbourhood of 100 °C. In other respects the structure of these materials is comparable to that of a rubber, and on heating to about 160 °C they acquire rubber-like properties.

1.7. Crystallization in raw rubber

It has been known for a long time that if rubber is maintained at a low temperature (e.g. 0 °C or lower) it gradually crystallizes. Unlike the glass transition, the change to the crystalline state is a *first-order* transition, and corresponds to a change of structure. As with other materials, crystallization in rubber is accompanied by an increase of density and by the release of latent heat. The presence of a crystalline phase, however, is most unambiguously revealed by the change in the X-ray diffraction pattern, which in the crystalline rubber contains a series of sharp rings, representing reflections from specific crystal planes, in addition to a weaker diffuse background arising from the residual disordered or amorphous component.

The process of crystallization may conveniently be followed by observations of the accompanying changes in specific volume. An example, taken from the early work of Bekkedahl (1934), is reproduced in Fig. 1.6. The characteristic S-shaped curve may be interpreted in terms of the processes of nucleation and crystal growth. As

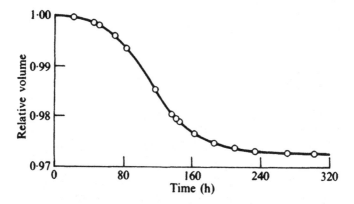

FIG. 1.6. Growth of crystallization in raw rubber at 0 °C, as measured by volume changes. (Bekkedahl 1934.)

in other processes involving the development of a new phase, crystallization takes place preferentially about existing nuclei, which may be thought of as arising initially by the random association of neighbouring segments of chains to form an ordered array. The increase in the rate of crystallization in the early stages is accounted for by the increasing size of the nuclei, and possibly also by an increase in the number of nuclei present. The final reduction in rate is due to the mutual interference between expanding nuclei, and the difficulty of disentangling the remaining amorphous chain segments so as to draw them into one or other of these centres of growth.

This general model accounts also for the effect of temperature on the rate of crystallization. This rate, as shown in Fig. 1.7, at first increases as the temperature is reduced, owing to the higher probability of formation of a nucleus. With further reduction in temperature, however, the decreasing mobility of chain segments begins to take effect, with the result that the rate of crystallization passes

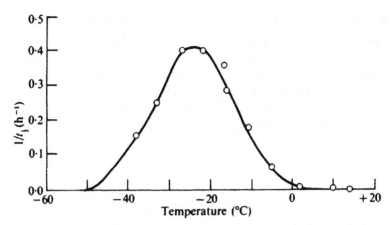

FIG. 1.7. Rate of crystallization $(1/t_{\frac{1}{2}})$ as function of temperature. $t_{\frac{1}{2}}$ = time for half final amount of crystallization to occur. (Wood 1946.)

through a maximum and ultimately falls to a very low value as the glass transition temperature is approached.

Detailed examination of the X-ray diffraction pattern of the crystallites shows the axes of the chains to lie parallel to one another in the unit cell. From the unit-cell dimensions the density of the crystal is estimated at $1 \cdot 00$ g cm^{-3} (Bunn 1942), which is 10 per cent higher than the density of the amorphous rubber ($0 \cdot 91$ g cm^{-3}). The maximum increase in density obtained on crystallizing rubber (Fig. 1.6) is well below this figure, and indicates that only about 27 per cent of the material is in the crystalline phase, the remainder being still in the amorphous condition. The picture which emerges is therefore of an assembly of crystallites formed by the three-dimensional ordering of chain segments interspersed within a continuous matrix of disordered, non-crystalline material (Fig. 1.8(a)).

The presence of an amorphous component accounts for the fact that crystalline rubber still shows a second-order transition (Fig. 1.3), though the change in expansivity at the transition temperature is correspondingly reduced in magnitude. Fig. 1.3 shows the crystalline state to be stable so long as the melting temperature is not approached. The temperature of melting of the crystallites is found to depend on the temperature at which the crystallization has taken place; for crystallization at 0 °C melting starts at about 6 °C and is complete at 16 °C.

Because of the reinforcing effect of the crystallites, crystalline rubber is considerably harder (by a factor of about 100) and less extensible than amorphous rubber, though it is still flexible, owing to the residual amorphous (rubber-like) component of the structure. It also has a yellowish waxy appearance, due to the reflection of light from clusters of crystals known as spherulites. (The individual crystallites have dimensions much smaller than the wavelength of light, i.e. 10–100 nm, and hence do not produce reflection.)

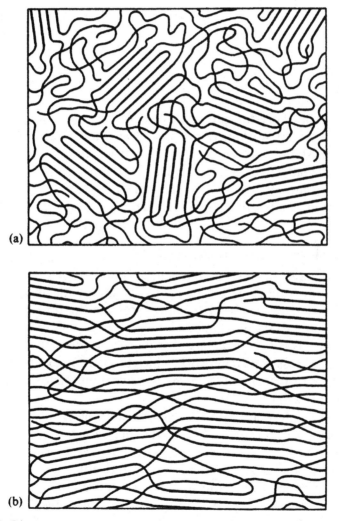

(a)

(b)

Fig. 1.8. Diagrammatic representation of structure of rubber crystallized in (a) the unstrained state and (b) the strained state.

The typical crystalline polymers—polythene, nylon, etc.—have structures and mechanical properties essentially similar to those of crystalline rubber, with the difference that their crystal melting points are considerably higher, so that they remain stable under ordinary conditions of use.

It should be emphasized that the representation of the structure given in Fig. 1.8 is undoubtedly greatly oversimplified. Our views on this subject have been greatly modified by the demonstration that single crystals of polymers such as polythene obtained by crystallization from dilute solution have the form of thin platelets or *lamellae*, which are produced by the regular folding backwards and forwards of single polymer chains, the chain axes being roughly perpendicular to the plane of the lamella. There is considerable evidence that chain-folded lamellae also form the basic structural entities in polymers crystallized from the melt (Geil 1963). A lamellar type of crystallization has also been identified in rubber by Andrews and co-workers (Andrews, Owen, and Singh 1971). The precise crystalline morphology in any particular case depends not only on the nature of the polymer but also on the detailed structural characteristics, such as molecular weight and molecular weight distribution, chain branching, etc., and the conditions of crystallization.

1.8. Crystallization in the stretched state

The general features of the crystallization process which are found in raw rubber occur also, with certain modifications, in the vulcanized material. The chief difference is in the time-scale, vulcanized rubber crystallizing much more slowly under comparable conditions. This difference arises from the presence in vulcanized rubber of a cross-linked network, which to a certain extent impedes the relative displacement of chain segments required for the formation and growth of nuclei.

Another difference is in the phenomenon of crystallization by stretching. Vulcanized rubber in the unstrained state does not crystallize at an appreciable rate at room temperature, but the application of a high extension produces immediate crystallization. X-ray studies show the pattern of this crystallization to be different from that for normal crystallization in the unstrained state, the crystallites showing a preferred orientation in the direction of the extension. The amount of crystallization increases rapidly with

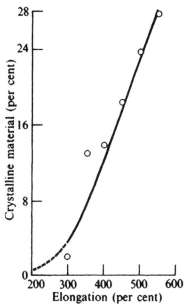

FIG. 1.9. Percentage crystallinity in vulcanized rubber as function of elongation. (Goppel 1949.)

increasing extension (Fig. 1.9). On retraction from the stretched state the original amorphous condition is simultaneously restored.

This phenomenon of spontaneous crystallization on extension appears at first sight extremely difficult to understand. Studies on raw rubber, however, have greatly assisted in obtaining a solution to this problem. Raw rubber has the advantage that the process of crystallization can be readily followed at all states of strain, including the unstrained state, and not only at high extensions. Furthermore, when crystallized in the stretched state, raw rubber does not retract to the unstrained state on removal of the stretching force, there being no cross-linked elastic network to provide the necessary restoring force.

Fig. 1.10 shows the density changes which accompany the crystallization of raw rubber at 0 °C in the unstrained state, and at various extensions up to 700 per cent (Treloar 1941). From these and similar observations it is concluded that there is no distinction in principle between crystallization in the unstrained state and strain-induced crystallization, the only significant difference being in the *rate* at which the process takes place, which is highly sensitive to the degree of orientation or preferential alignment of the chains. At the

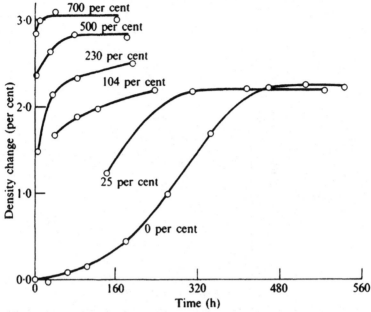

FIG. 1.10. Density changes in raw rubber crystallized at various extensions (values given against curves) at 0 °C.

highest extensions (~700 per cent) the process takes place so rapidly that its earlier stages escape observation.

Mechanical properties of stretched crystalline rubber

The structure of raw rubber after crystallization in the stretched state may be represented as a system of roughly parallel crystallites with interconnecting amorphous material, as in Fig. 1.8(b). This type of structure is analogous to that of a fibre, such as, for example, nylon or Terylene. The similarity in structure extends also to the mechanical properties. Stretched crystalline rubber is highly aniso-tropic in mechanical properties, possessing a high strength in the direction of the extension but very low strength in the transverse direction. These 'fibrous' properties of highly stretched raw rubber were first demonstrated by Hock (1925), who showed that on cooling in liquid air and hammering it split up into bundles of fibrils. This effect he attributed to crystallization, which was first conclu-sively demonstrated independently by the direct X-ray observa-tions of Katz (1925).

There is considerable evidence (Gee 1947; Morrell and Stern 1953) that crystallization also plays a significant part in enhancing the strength of vulcanized rubber. Many of the synthetic rubbers, particularly those formed by the random co-polymerization of two different monomer units, such as butadiene–styrene rubber (Table 1.1, p. 5), do not have a regular sequence of chain atoms and hence are incapable of forming a crystal lattice. Such rubbers, after vulcanization, generally have very much lower strengths than natural rubber vulcanizates, though from the practical standpoint this difference can be largely overcome by suitable compounding techniques, e.g. by the incorporation of a reinforcing filler such as carbon black.

INTERNAL ENERGY AND ENTROPY
CHANGES ON DEFORMATION

2.1. Stress–temperature relations

IN the preceding chapter attention was drawn to the peculiar thermoelastic phenomena encountered with rubber, and the significance of these phenomena in relation to the kinetic theory of rubber elasticity proposed by Meyer and his associates was emphasized. In the present chapter we shall examine in closer detail the underlying thermodynamic relations involved in the study of these thermoelastic phenomena and the way in which they may be applied experimentally to obtain quantitative information bearing on the question of the mechanism of the deformation process.

At this stage the treatment will be limited to the more elementary aspects of the subject which are required in order to establish a basis for the development of the statistical theory and for the proper understanding of its physical significance. The more extensive and elaborate later developments, which involve various refinements in matters of detail, must necessarily be deferred to a later chapter (Chapter 13).

The kinetic theory in its simplest form attributes the elasticity of rubber to changes in the conformations of a system of long-chain molecules in passing from the unstrained to the strained state. Such changes are associated with changes in the configurational entropy of the system, the internal energy being considered to remain unchanged. For any mechanism of this kind (i.e. involving only changes of entropy) it may be shown (as below) that the stretching force, for a given state of strain, should be proportional to the absolute temperature. The early experiments of Meyer and Ferri (1935) showed this conclusion to be substantially fulfilled, over a wide range of temperature, provided that the extension was sufficiently large (Fig. 2.1). In the region of lower strains, however, the behaviour appeared to be anomalous, the force increasing less rapidly than would be expected theoretically, or even actually *decreasing* as the temperature was raised. These observations have

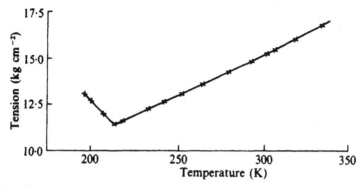

FIG. 2.1. Force at constant length as function of absolute temperature. Extension 350 per cent. (From Meyer and Ferri 1935.)

since been repeated by many other workers, notably by Anthony, Caston, and Guth (1942), Wood and Roth (1944) and, to quote a more recent example, Shen, McQuarrie, and Jackson (1967), with precisely similar results. The data presented in Fig. 2.2, reproduced from the work of Anthony *et al.* (1942), may be taken as typical of

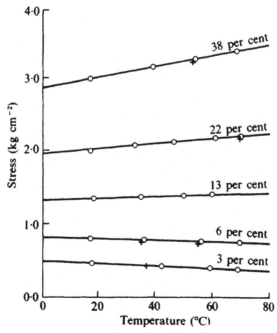

FIG. 2.2. Force at constant length as function of temperature. Elongations as indicated. (Anthony, Caston, and Guth 1942.)

the behaviour in the low-strain region. It will be seen that the reversal in slope of the stress–temperature plots occurs at an extension of about 10 per cent—the so-called *thermoelastic inversion point.*

The explanation of these peculiar effects is in fact very simple, and was correctly given by both Meyer and Ferri and by Anthony, Caston, and Guth. They arise from the normal thermal expansivity of the unstrained rubber, as a result of which the unstrained length of the sample varies with change in temperature. As a result, an increase of temperature at *constant length* involves a reduction of the *strain* or relative extension, and this in itself will cause a reduction of the applied force. Thus even though the force at a given value of *strain* may increase with increasing temperature, this increase may be more than counterbalanced by the associated reduction in strain. This will be seen more clearly from Figs 2.3 and 2.4. If the elongation is calculated on the basis of the unstrained

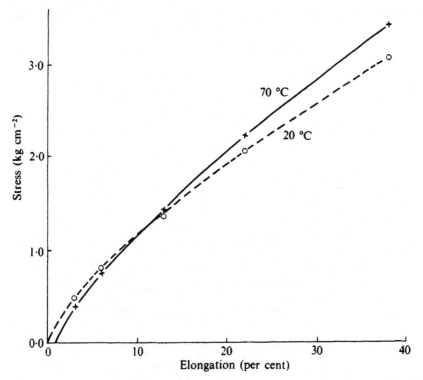

FIG. 2.3. Force at 20 °C and 70 °C plotted against extension calculated on unstrained length at 20 °C. (Anthony, Caston, and Guth 1942.)

length at 20 °C, the origin of the force–elongation curve for 70 °C is displaced; hence although the slope is higher, the force is initially lower, the two curves intersecting at about 10 per cent strain (Fig. 2.3). If, however, the extensions are calculated on the basis of the unstrained length at the temperature of operation, the curves for different temperatures all pass through a single origin, and the values of force, for any given value of strain, are almost exactly proportional to absolute temperature (Fig. 2.4). This adjustment, therefore, eliminates the thermoelastic inversion effect.

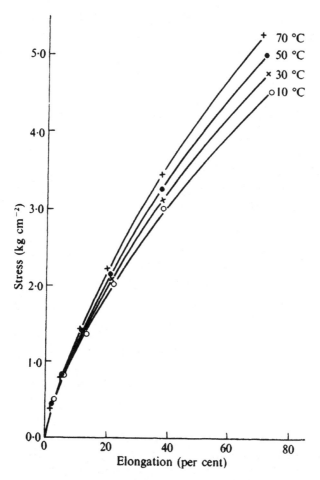

FIG. 2.4. Force at various temperatures plotted against extension calculated on unstrained length at temperature of measurement. (Anthony, Caston, and Guth 1942.)

2.2. Thermodynamic analysis

For the more explicit examination of these thermoelastic phenomena it is necessary to develop relations between force, length, and temperature on the one hand and the thermodynamic quantities, internal energy and entropy, on the other. The required relations are obtainable directly from the first and second laws of thermodynamics. From the first law of thermodynamics the change in internal energy dU in any process is given by

$$dU = dQ + dW, \qquad (2.1)$$

where dQ and dW are respectively the heat absorbed by the system and the work done on it by the external forces. The second law defines the entropy change dS in a *reversible* process by the relation

$$T\, dS = dQ, \qquad (2.2)$$

and hence from (2.1), we have for a reversible process

$$dU = T\, dS + dW. \qquad (2.3)$$

In discussing the equilibrium of a system which is subject to reversible changes (e.g. elastic deformations) it is convenient to introduce the *Helmholtz free energy A*, defined by the relation

$$A = U - TS. \qquad (2.4)$$

For a change taking place at *constant temperature* we have then

$$dA = dU - T\, dS. \qquad (2.5)$$

Combining this equation with (2.3) we obtain the standard thermodynamic result

$$dA = dW \quad \text{(constant } T\text{)}, \qquad (2.6)$$

which signifies that in a reversible isothermal process the change in Helmholtz free energy is equal to the work done on the system by the applied forces.

In most thermodynamic textbooks the subject is subsequently developed with particular reference to gases and liquids, for which the significant variables include pressure p and volume V. The work done on the system in a small displacement is then written as

$$dW = -p\, dV.$$

In discussing problems related to the elasticity of solids, on the other hand, we are concerned primarily with the work done by the applied stress, corresponding, for example, to a tensile force f acting on a specimen of length l, in which case the work done in a small displacement is

$$dW = f \, dl. \tag{2.7}$$

When in addition a hydrostatic pressure (e.g. atmospheric pressure) is also present the total work done by the applied forces becomes

$$dW = f \, dl - p \, dV. \tag{2.7a}$$

For strict accuracy it is necessary to take account of both terms on the right-hand side of (2.7a). But in the case of rubbers, the volume change dV is usually very small, and if p is the atmospheric pressure the term $p \, dV$ is less than $f \, dl$ by a factor of 10^{-3} or 10^{-4}. As a first approximation we may therefore neglect this term and use eqn (2.7), which is strictly accurate only at zero applied pressure or under constant-volume conditions, in place of the exact expression (2.7a) (the more accurate analysis, not involving this assumption, is referred to in § 2.4). By making use of eqns (2.6) and (2.7) the tension may then be expressed in the form

$$f = \left(\frac{\partial W}{\partial l}\right)_T = \left(\frac{\partial A}{\partial l}\right)_T, \tag{2.8}$$

which shows that the tensile force is equal to the change in Helmholtz free energy per unit increase in length of the specimen.

The significance of the important relation (2.8) may be better appreciated by reference to Fig. 2.5, which represents diagrammatically the variation of Helmholtz free energy for an elastic body as a function of its length l. The unstressed state is such that the Helmholtz free energy is a minimum, so that $(\partial A/\partial l)_T = 0$ when $l = l_0$. If l is greater than l_0, $(\partial A/\partial l)_T$ is positive, corresponding to a tensile force, while if l is less than l_0, $(\partial A/\partial l)_T$ is negative, corresponding to a compressive force. (It should be noted, however, that the force–deformation relation will not in general be linear, except for very small strains.)

The tension, like the free energy, may be expressed as the sum of two terms (from eqn (2.5)), thus

$$f = \left(\frac{\partial A}{\partial l}\right)_T = \left(\frac{\partial U}{\partial l}\right)_T - T\left(\frac{\partial S}{\partial l}\right)_T, \tag{2.9}$$

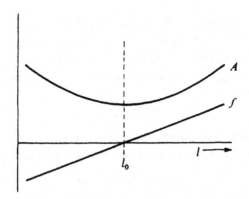

FIG. 2.5. Dependence of Helmholtz free energy A and force f on length l for specimen subjected to uniaxial extension or compression. l_0 = unstrained length.

of which the first represents the change in internal energy and the second the change in entropy, per unit increase in length. The second term is related to the temperature coefficient of tension, as will now be shown.

Writing eqn (2.4) in differential form, we have for any change (i.e. not necessarily isothermal)

$$dA = dU - T\,dS - S\,dT,$$

whilst from (2.3) and (2.7)

$$dU = f\,dl + T\,dS.$$

Combination of these two equations gives

$$dA = f\,dl - S\,dT.$$

Hence, by partial differentiation,

$$\left(\frac{\partial A}{\partial l}\right)_T = f; \qquad \left(\frac{\partial A}{\partial T}\right)_l = -S. \tag{2.10}$$

By a well-known property of partial differentials

$$\frac{\partial}{\partial l}\left(\frac{\partial A}{\partial T}\right)_l = \frac{\partial}{\partial T}\left(\frac{\partial A}{\partial l}\right)_T,$$

and hence, from eqns (2.10),

$$\left(\frac{\partial S}{\partial l}\right)_T = -\left(\frac{\partial f}{\partial T}\right)_l. \tag{2.11}$$

Eqn (2.11) gives the entropy change per unit extension in terms of the measurable quantity $(\partial f/\partial T)_l$, the temperature coefficient of tension at constant length. Insertion of this relation in (2.9) gives for the corresponding internal energy change

$$\left(\frac{\partial U}{\partial l}\right)_T = f - T\left(\frac{\partial f}{\partial T}\right)_l \; . \qquad (2.12)$$

The relations (2.11) and (2.12) are of fundamental importance in rubber elasticity, since they provide a direct means of determining experimentally both the internal energy and entropy changes accompanying a deformation. Their application may be illustrated by reference to Fig. 2.6, in which the curve CC′ represents the

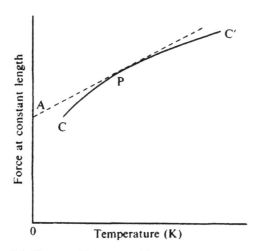

FIG. 2.6. Slope and intercept of force–temperature curve.

variation with temperature of the force at constant length, which may or may not be linear. The slope of this curve at a point P is $(\partial f/\partial T)_l$, which from eqn (2.11) gives the entropy change per unit extension $(\partial S/\partial l)_T$ for an isothermal extension at the temperature T. In a corresponding way, the intercept 0A of the tangent to the curve at P on the vertical axis $T = 0$ is $f - T(\partial f/\partial T)_l$, which by eqn (2.12) is equal to the change of internal energy per unit extension.

Thus the internal energy and entropy contributions to the force at any given value of the extension are directly obtainable from the experimental force–temperature plot. In particular, if the force–temperature plot is linear (as in Fig. 2.1) both internal energy and entropy terms are independent of temperature. If in addition the force–temperature relation is represented by a straight line passing through the origin, the internal energy term is zero, i.e. the elastic force arises solely from the change in entropy on extension.

2.3. Application to experimental data

In applying eqns (2.11) and (2.12) to the derivation of internal energy and entropy changes on extension, difficulties may arise on account of the imperfect elasticity or reversibility of the rubber. The analysis presupposes that for any given strain and temperature the value of the applied force is uniquely determined. However, even when vulcanized, rubber suffers from some degree of irreversibility, as evidenced by stress–relaxation and creep effects, which cannot be

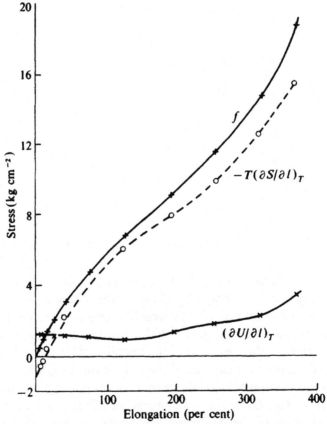

FIG. 2.7. Internal energy and entropy components of tensile force, as functions of extension. f = force, U = internal energy, and S = entropy. (Anthony, Caston and Guth 1942.)

entirely eliminated. It is found, however, that if for each value of the strain the rubber is held at a relatively high temperature (e.g. 70–100 °C) for a sufficient time to allow the stress to relax to a substantially constant value, then effectively reversible stress–temperature plots are obtained for all variations of temperature not exceeding the original temperature of relaxation. This or some equivalent procedure has therefore been adopted in all experimental studies of this kind.

The application of eqns (2.11) and (2.12) to the stress–temperature data of Anthony *et al.* (1942), examples of which are given in Fig. 2.2, yielded the results shown in Fig. 2.7. It is seen that

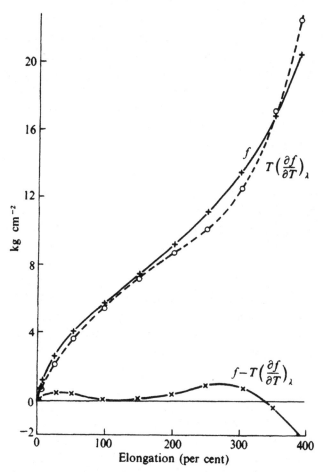

FIG. 2.8. Slope $T(\partial f/\partial T)_\lambda$ and intercept of force–temperature relation at constant extension ratio. (Anthony, Caston, and Guth 1942.)

for extensions exceeding 100 per cent the dominant component of the force is the entropy component $-T(\partial S/\partial l)_T$, but that there is in addition a significant internal energy contribution. As the extension is decreased this internal energy term becomes relatively more important, and in the region of small extensions (0–10 per cent) it is the major factor in the situation. Indeed, in this region it is hardly correct to refer to the internal energy and entropy *contributions* to the force, except in the purely formal sense, because the internal energy term actually *exceeds* the total force while the corresponding entropy term, being negative, is subtracted from it.

These peculiarities reflect the changes in slope of the force–temperature relations at constant length, and may similarly be eliminated by appropriate adjustment of the unstrained length to compensate for the thermal expansivity of the unstrained rubber. Application of this 'correction' yields the force–temperature relation at *constant extension ratio* λ, i.e. $(\partial f/\partial T)_\lambda$ rather than at constant length l. Values of this quantity derived from the same original data are given in Fig. 2.8, together with the corresponding 'intercept' at $T = 0$ K, namely, $f - T(\partial f/\partial T)_\lambda$. To the degree of accuracy obtainable, this intercept is approximately zero.

2.4. Interpretation of thermoelastic data

To understand the above results it is necessary to take account of the changes of volume which accompany the application of a stress. These volume changes arise from the hydrostatic component of the applied tensile stress, and are associated with the bulk compressibility of the material. To take account of such volume changes it is necessary to distinguish between extensions at constant pressure and at constant volume. The experimental stress–temperature relations are normally obtained at constant (i.e. atmospheric) pressure, and the corresponding entropy and internal energy components are expressed by the modified relations

$$\left(\frac{\partial S}{\partial l}\right)_{p,T} = -\left(\frac{\partial f}{\partial T}\right)_{p,l}; \quad \left(\frac{\partial U}{\partial l}\right)_{p,T} \approx \left(\frac{\partial H}{\partial l}\right)_{p,T} = f - T\left(\frac{\partial f}{\partial T}\right)_{p,l}, \quad (2.13)$$

in which H is the heat content ($U + pV$). In the present situation this does not differ significantly from the internal energy U.

The volume changes due to the applied stress, though small (i.e. of the order of 10^{-4}), nevertheless make a significant contribution to the internal energy. This contribution, however, arises from the forces *between* the molecules, and is in no way directly related to the deformation of the network of long-chain molecules, which is our primary concern. It would be physically more relevant to our present problem to consider a deformation at constant *volume*. The maintenance of constancy of volume during deformation would require the superposition of a hydrostatic pressure. Such an experiment would be difficult to carry out, and though it was eventually successfully performed in 1963 (cf. Chapter 13), this was not visualized as a practical possibility in the early days of the subject. However, both Elliott and Lippmann (1945) and Gee (1946a) developed general thermodynamic formulae which enabled the relations between the stress–temperature coefficients at constant volume and at constant pressure and the corresponding thermodynamic quantities $(\partial U/\partial l)_{V,T}$ and $(\partial U/\partial l)_{p,T}$ to be deduced. They also showed that it was possible to derive the internal energy change at constant volume from measurements at constant pressure by considering the stress–temperature coefficient at constant extension ratio λ (i.e. constant strain) rather than at constant length. The appropriate relation, which is only an approximation, is

$$\left(\frac{\partial U}{\partial l}\right)_{V,T} \simeq f - T\left(\frac{\partial f}{\partial T}\right)_{p,\lambda}. \tag{2.14}$$

This result provides an immediate interpretation of the data of Anthony *et al.* reproduced in Fig. 2.8. The interpretation is that the internal energy contribution to the force, at *constant volume*, is approximately zero, in agreement with the basic postulate of the kinetic theory.

This important conclusion implies that the observed internal energy changes which occur under the normal constant pressure conditions are due to the accompanying changes of volume. Pursuing this argument, Gee derived a quantitative relation between the internal energy and the change of volume on extension, from which he was able to predict the amount of the volume change to be expected.

The above conclusion does not apply at very high extensions, where the effects of crystallization become apparent. Anthony *et al.* employed a natural rubber vulcanizate containing 8 per cent of sulphur; this was chosen because of its good reversibility and absence of crystallinity on extension. Wood and Roth (1944), who made a similar study covering a wider range of extension, worked with a more normal low-sulphur vulcanizate. Their results showed an increasingly large *negative* internal energy component as the extension was increased from 200 per cent to 700 per cent (Fig. 2.9). That this was to be associated with crystallization was confirmed by

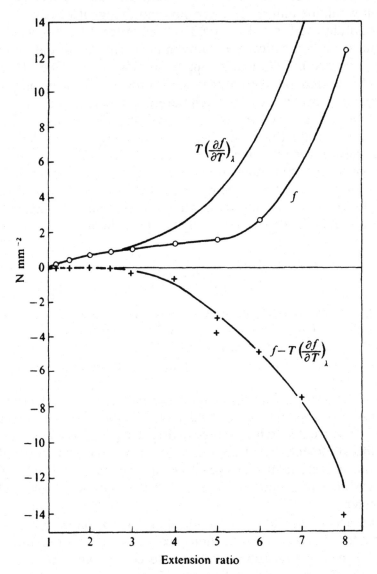

FIG. 2.9. Slope $T(\partial f/\partial T)_\lambda$ and intercept of force–temperature relation at constant extension ratio. (Wood and Roth 1944.)

the fact that in a similar study of the non-crystallizing styrene–butadiene rubber the internal energy changes were comparatively small (Roth and Wood 1944).

It is necessary to emphasize that the theoretical relation (2.14) is only an approximation. Its derivation involves the assumption that the rubber is isotropically compressible in the strained state, i.e. that

$$\left(\frac{\partial l}{\partial V}\right)_{f,T} = \frac{l}{3V}, \tag{2.15}$$

which, as Gee (1946a) fully appreciated, is strictly valid only in the limiting case of very small strains ($\lambda \rightarrow 1$), and is likely to become increasingly inexact with increasing strain, particularly when crystallization occurs. A more realistic treatment, taking into account the anisotropy of compressibility in the strained state, cannot be derived on the basis of purely thermodynamic reasoning but necessitates the introduction of a specific physical or molecular model of the structure. Such a model has in fact been developed by Flory (1961), and is fully discussed in Chapter 13. This shows that the conclusion drawn from the original studies by Gee, though broadly correct, require some modification in detail. In particular, it shows that, while the volume changes are responsible for the *major* contribution to the internal energy changes which accompany an extension at constant pressure, they are not, in general, the sole factor in the situation, there being a further contribution arising from energetic interactions within the single molecule. This contribution arises even when the deformation is carried out under constant-volume conditions.

2.5. Thermal effects of extension

Reference was made in Chapter 1 to the observations of Joule on the heat of extension, as revealed by the rise of temperature (Fig. 2.10). The explanation of this effect is very simple, and follows directly from the basic concept of the kinetic theory, namely, that the deformation of a rubber (at constant temperature) is associated with a reduction of entropy, with no change in internal energy. Putting $dU = 0$ in eqns (2.1) we obtain

$$dW = -dQ \quad \text{(constant } T\text{)}. \tag{2.16}$$

The work done by the stretching force being necessarily positive, it follows that dQ, the heat *absorbed*, is negative, i.e. that heat is

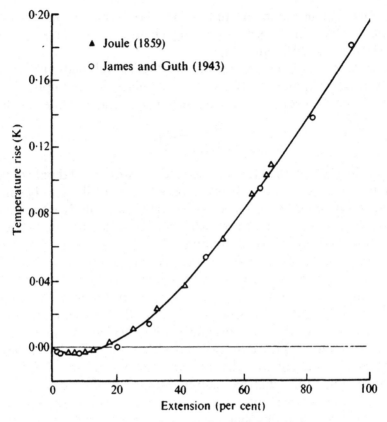

FIG. 2.10. Temperature rise in adiabatic extension.

evolved on extension. The amount of the heat evolution is exactly equal to the work done on the rubber by the applied force.

The physical meaning of this conclusion is as follows. The internal energy of the rubber is purely kinetic, and arises from the thermal agitation of the constituent atoms of the chains. This energy is a function only of the temperature, and is independent of the conformations of the chains, i.e. of the state of strain. Since the internal energy does not change, the work performed by the applied force in an isothermal deformation must be balanced by the emission of an equivalent amount of heat.

If this heat is not emitted but is retained, as in an adiabatic extension, the energy supplied by the applied force is used to increase the molecular agitation, i.e. to produce a rise in temperature. The temperature rise is therefore $-\mathrm{d}Q/c_l$, where c_l is the heat capacity (specific heat) at constant length and $-dQ$ is the equivalent heat emission in an incremental extension dl at constant temperature. The total temperature rise ΔT in an extension from the unstrained length l_0 to the final length l is obtained by integration of $-dQ$ with respect to l. Since the process is reversible we may put $dQ = T\,dS$ and hence obtain

$$\Delta T = -\frac{T}{c_l} \int_{l_0}^{l} \left(\frac{\partial S}{\partial l}\right)_T dl, \qquad (2.17)$$

it being assumed that c_l does not vary appreciably with length. This equation provides a direct link between the heat of extension and the entropy changes on extension, as determined from the stress-temperature relations.

Fig. 2.10 gives, in addition to Joule's original data, the later results given by James and Guth (1943) on the temperature rise in an adiabatic extension. These are seen to be of precisely similar form, though the quantitative agreement is fortuitous, since different samples of vulcanized rubber will in general not be identical in physical properties. The initial *cooling* effect corresponds to the *positive* entropy of deformation, which we have already seen to be characteristic of the low-extension region. Since the quantity ΔT involves an integration, the minimum in the ΔT curve should correspond to the thermoelastic inversion point, at which $(\partial S/\partial l)_T$ is zero. Experimentally, the minimum is seen to occur at an extension somewhat below 10 per cent (7 per cent to be precise), which is consistent with the range of observed values of the inversion point. There is thus a complete parallel between the experimental temperature changes on extension and the thermoelastic data. This conclusion is further reinforced by more recent measurements of the heat of extension, discussed in Chapter 13.

In the region of high strains the observed thermal effects are complicated by the incidence of crystallization which gives rise to the evolution of a 'latent heat'. This will be apparent from Fig. 2.11, in which the temperature rise is seen to increase steeply in the region beyond 300 per cent extension. The lack of reversibility between extension and retraction measurements is a further indication of crystallization, which tends to persist as the extension is reduced (cf. Fig. 9.7, p. 190). These effects were not observed in the

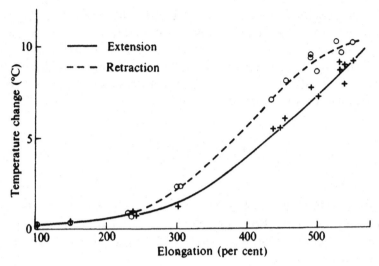

FIG. 2.11. Temperature changes in adiabatic extension (or retraction). (Dart, Anthony, and Guth 1942.)

case of a rubber such as butadiene–acrylonitrile which does not crystallize on extension.

2.6. Conclusion

The consideration of the thermodynamics of the process of extension is fundamental to the development of the statistical theory, as given in subsequent chapters. On the simplest basis, as we have already noted, this theory postulates that the deformation is accompanied by a reduction of entropy, without change in internal energy. Analysis of the available experimental data shows this assumption to be approximately justified at moderately large strains, but to be very far from the truth at small strains, in the region of 10 per cent extension or below. However, the major part (though not necessarily the whole) of the observed internal energy changes may be accounted for in terms of the associated changes of volume which accompany the application of the stress. These volume changes depend on the finite compressibility of the rubber, which is not taken into account in the elementary presentation of the statistical theory, since it is not directly related to the mechanism of the deformation process. Further complications may arise in the region of very high extensions, as a result of crystallization, which is associated with a relatively large reduction in volume, and a correspondingly large reduction in internal energy. When due allowance is made for these secondary effects, however, it is clear that the basic thermoelastic and thermal observations are consistent with the postulates of the statistical theory and provide a direct proof of the underlying kinetic mechanism of the deformation process.

THE ELASTICITY OF
LONG-CHAIN MOLECULES

3.1. Statistical properties of long-chain molecules

THE preceding chapters have set out the basic concepts of the
kinetic theory of rubber elasticity and discussed in general thermodynamic terms some of its most characteristic consequences. We
have seen that in its dependence on the changes of conformation of
long-chain molecules the elasticity of rubber stands in striking
contrast, both physically and thermodynamically, to the elasticity of
an ordinary solid such as a crystal or a glass. We now proceed to the
next stage in the development of the kinetic theory, which consists
in the quantitative derivation of the actual mechanical properties of
a rubber in terms of its molecular constitution. This development
involves two essentially separate issues, which have already been
hinted at in Chapter 1. These are first, the treatment of the statistical
properties of the single long-chain molecule in terms of its geometrical structure, and secondly, the application of this treatment to the
problem of the network of long-chain molecules corresponding to a
cross-linked or vulcanized rubber. The first of these issues is the
subject of the present chapter.

The study of the statistical properties of the long-chain molecule
is a complex one, and may be carried to various stages of refinement. In the present context we shall be concerned primarily with
those statistical properties which are common to all rubber-like
molecules, and which may conveniently be dealt with in terms of an
idealized chain of freely rotating links. The statistical properties of
such an idealized model involve geometrical considerations only,
and may be dealt with quite simply. The more realistic treatment of
an actual molecule, however, requires that the effects of energetic
interactions between different segments of the same chain should be
taken into account. This is a problem of considerably greater
complexity, which falls outside the scope of the present work.

Even with the simplest model of the chain the degree of
mathematical difficulty depends on the range of extension considered. If only moderate extensions are involved a very simple

statistical treatment, known as the Gaussian treatment, is sufficient. But for the higher range of extensions this treatment becomes increasingly inadequate as the distance between the two ends of the chain approaches the fully extended length, and a more elaborate non-Gaussian treatment has to be used. The Gaussian treatment for the single chain will be developed in the present chapter, and its application to the problem of the network in the one which follows. The development of the non-Gaussian theory will be presented in Chapter 6.

3.2. Statistical form of long-chain molecule

The statistical form of the long-chain molecule may be illustrated by considering an idealized model of the polymethylene or paraffinic type of chain $(CH_2)_n$, in which the angle between successive bonds (i.e. the valence angle) is fixed but complete freedom of rotation of any given bond with respect to adjacent bonds in the chain is allowed. This is illustrated in Fig. 3.1, in which the first two

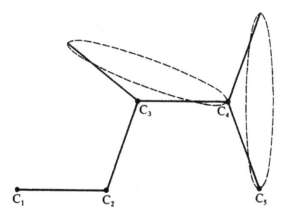

FIG. 3.1. Rotation about bonds in paraffin-type molecule.

bonds C_1C_2 and C_2C_3 are represented as lying in the plane of the paper. The third bond, C_3C_4, will in general not lie in this plane but will rotate in a random manner about the bond C_2C_3 as axis. Similarly, C_4C_5 will rotate about C_3C_4, and so on. The chain will thus take up an irregular or randomly kinked form (Fig. 3.2(b)) in which the distance between its ends is very much less than that corresponding to the outstretched or planar zig-zag form (Fig. 3.2(a)). The actual conformation will be subject to continual fluctuation due

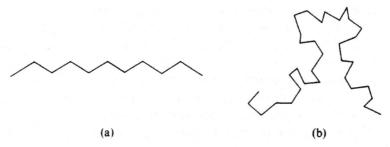

(a)   (b)

FIG. 3.2. (a) Planar zig-zag; (b) randomly kinked chain.

to thermal agitation and hence cannot be defined explicitly, but it is possible to specify some of the properties of the system in statistical terms, or in terms of certain average values. Thus, for example, it may be shown that the root-mean-square (r.m.s.) value of the distance r between the ends of a chain containing n bonds (where n is a large number) is given by

$$(\overline{r^2})^{\frac{1}{2}} = ln^{\frac{1}{2}}\left(\frac{1+\cos\theta}{1-\cos\theta}\right)^{\frac{1}{2}}, \tag{3.1}$$

where l is the bond length and θ the supplement to the valence angle. Taking the valence angle to be $109\frac{1}{2}°$, $\theta = 70\frac{1}{2}°$, and therefore $\cos\theta = \frac{1}{3}$, hence for this case

$$(\overline{r^2})^{\frac{1}{2}} = l(2n)^{\frac{1}{2}}. \tag{3.1a}$$

This result illustrates the quite general conclusion that the mean (or r.m.s.) dimensions of *any* chain of this kind are proportional to the *square root* of the number of bonds or links which it contains.

The general form of the idealized freely rotating chain is more clearly indicated in Fig. 3.3, which represents an actual wire model of a polymethylene chain containing 1000 links. In the construction of this model the links were set at the required valence angle, but the position of each successive link in the circle of rotation was chosen at random by the throw of a die. This gave an equal probability for each of six equally spaced positions in the circle, which may be regarded as a sufficiently close approximation to complete randomness. The photographic reproduction in Fig. 3.3 is, of course, equivalent to a two-dimensional projection of the actual three-dimensional form. The particular conformation obtained is just one

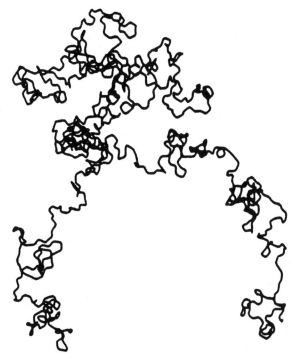

FIG. 3.3. Form of 1000-link polymethylene chain according to the statistical theory.

of a very large number of equally probable forms (6^{998} to be precise) which might have been produced by this method of construction, but it happens to be a fairly 'average' sample, the end-to-end distance being within about 10 per cent of the calculated r.m.s. value (eqn (3.1a)).

Polyisoprene chains

The calculation of the mean-square length of a molecule is not limited to the case where all the bonds and valence angles are equal. A general method of treating more complicated structures has been worked out by Wall (1943) and applied to the polyisoprene chain (Fig. 1.2, p. 4) in which there is one double bond per four chain carbon atoms. The two single bonds adjacent to the double bond are fixed in a plane; apart from this restriction, freedom of rotation about single bonds is assumed. Wall considered both natural rubber, in which the single bonds have the *cis*-configuration with respect to the double bond, and gutta-percha, which is otherwise similar to natural rubber, but in which the *trans*-configuration occurs. For the former he obtained, on the basis of accepted values of bond lengths and valence angles, the numerical result

$$(\overline{r^2})^{\frac{1}{2}} = 0 \cdot 201 n^{\frac{1}{2}} \text{ nm}, \tag{3.2}$$

where n (assumed large) is the total number of bonds in the chain (including double bonds).

3.3. The randomly jointed chain

From the preceding examples it may be inferred that the general form of the long-chain molecule, as depicted in Fig. 3.3, is independent of the precise geometry of the chain, provided only that the number of bonds about which free (or relatively free) rotation can occur is sufficiently large. The particular geometry of the chain affects only the mean-square length, or average dimensions, of the chain, not its general form. For the development of the statistical theory in general mathematical terms it is convenient to make use of an idealized model of the simplest possible kind, which does not correspond directly to any actual molecular structure. This consists of a chain of n links of equal length l, in which the direction in space of any link is entirely random and bears no relation to the direction of any other link in the chain. Such a randomly jointed chain automatically excludes valence angle or other restrictions on the freedom of motion of neighbouring links.

In order to define the statistical properties of the randomly jointed chain we consider one end A to be fixed at the origin of a Cartesian coordinate system Ox, Oy, Oz and allow the other end B to move in a random manner throughout the available space (Fig. 3.4). However, though the motion is random, all positions of B are not equally probable, and for any particular position P, having coordinates (x, y, z), there will be an associated probability that the end B shall be located within a small volume element $d\tau$ in the vicinity of the point P, which for convenience may be taken as a rectangular block of volume $dx\, dy\, dz$. The derivation of this probability requires the evaluation of the relative numbers of configurations or conformations of the chain which are consistent with different positions of the point P, the probability of any particular position being taken as proportional to the corresponding number of conformations. The solution to this problem, which has been

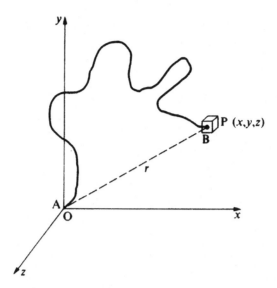

FIG. 3.4. The statistically kinked chain. Specification of probability that the end should fall in the volume element $d\tau (= dx\, dy\, dz)$.

given by Kuhn (1934, 1936) and by Guth and Mark (1934) is represented by the equation

$$p(x, y, z)\, dx\, dy\, dz = \frac{b^3}{\pi^{\frac{3}{2}}} \exp\{-b^2(x^2 + y^2 + z^2)\}\, dx\, dy\, dz, \qquad (3.3)$$

where

$$b^2 = 3/2nl^2. \qquad (3.3a)$$

This formula gives the probability that the components of the vector r representing the end-to-end distance for the chain shall lie within the intervals x to $x + dx$, y to $y + dy$, and z to $z + dz$, respectively. This probability is expressed as the product of the probability function $p(x, y, z)$ or *probability per unit volume* (sometimes referred to as the *probability density*), and the size of the volume element considered, which in this case is $dx\, dy\, dz$.

The formula (3.3) is of fundamental importance in the statistical theory of rubber elasticity. Its form is that of the Gaussian error function, which is of frequent occurrence in statistical problems concerned with the superposition of random effects. It has a number of interesting properties, some of which are considered below. It is important to note, however, that this form is only approximate, since its derivation involves the assumption that the distance r

between the ends of the chain is not comparable with the maximum or fully extended length nl of the chain, i.e. that

$$r \ll nl. \tag{3.4}$$

The precise region over which this approximation may be regarded as valid will depend on the accuracy required in any particular case; for higher extensions of the chain the more accurate non-Gaussian theory referred to earlier is required.

3.4. Properties of Gaussian functions

We note first that the function (3.3) is spherically symmetrical, for on putting $x^2 + y^2 + z^2 = r^2$ we obtain the probability density in the form

$$p(x, y, z) = (b^3/\pi^{\frac{3}{2}}) \exp\{-b^2(x^2 + y^2 + z^2)\} = (b^3/\pi^{\frac{3}{2}}) \exp(-b^2 r^2), \tag{3.5}$$

which is a function of r only. This result is, of course, to be expected, and merely implies that all directions of the vector OP (Fig. 3.4) are equally probable. Furthermore, the probability density $p(x, y, z)$ is a maximum when $r = 0$, that is when the two ends of the chain are coincident, and falls continuously as r increases. This means that if one end of the chain is fixed at the point O, the *most probable position* of the other end is *at the same point* O.

A second very important property of the Gaussian function (3.3) is that it can be represented as the product of three independent probabilities in each of the separate coordinates, i.e.

$$p(x, y, z)\, dx\, dy\, dz = \frac{b}{\pi^{\frac{1}{2}}} \exp(-b^2 x^2)\, dx \times$$

$$\times \frac{b}{\pi^{\frac{1}{2}}} \exp(-b^2 y^2)\, dy \times \frac{b}{\pi^{\frac{1}{2}}} \exp(-b^2 z^2)\, dz. \tag{3.6}$$

This implies that the probability that the chain shall have a particular component of length in, say, the x direction is completely independent of its components of length in the y and z directions. (This property of the chain is valid only in the Gaussian region of extension, defined by the condition (3.4). In the extreme case, if the chain is stretched to its fullest extent along the x-axis, the probability of a finite value of y or z is necessarily zero.)

The separability of the component probabilities enables us to integrate over all values of y and z, so as to obtain the total probability of a given component of length in the x direction, regardless of the values of y or z. Since

$$\int_{-\infty}^{+\infty} \exp\left(-b^2 y^2\right) dy = \int_{-\infty}^{+\infty} \exp\left(-b^2 z^2\right) dz = \frac{\pi^{\frac{1}{2}}}{b}, \qquad (3.7)$$

the result is

$$p(x)\, dx = (b/\pi^{\frac{1}{2}}) \exp\left(-b^2 x^2\right) dx. \qquad (3.8)$$

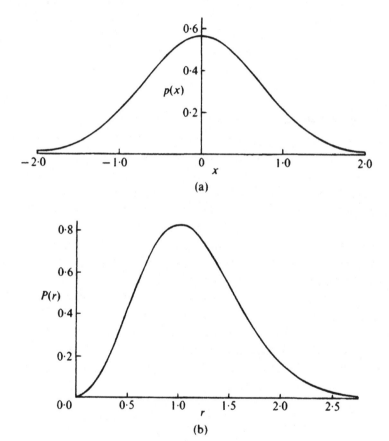

FIG. 3.5. Distribution functions: (a) $p(x) = \text{constant} \times \exp\left(-b^2 x^2\right)$; (b) $P(r) = \text{constant} \times r^2 \exp\left(-b^2 r^2\right)$.

The form of this function is shown in Fig. 3.5(a), from which it is seen that the probability of a component of length between x and $x + dx$ is given by the area of an elementary strip of width dx and height $p(x)$. The constant $b/\pi^{\frac{1}{2}}$, called the *normalizing factor*, is chosen so that the total probability of any value of x is unity, i.e.

$$\int_{-\infty}^{+\infty} p(x)\, dx = 1, \tag{3.8a}$$

in accordance with (3.7).

The function (3.8), like (3.3), contains only a single parameter b, which is related to the number of links in the chain and the link length l through (3.3a). This means that these functions have the same general form for all randomly jointed chains, but differ in breadth, according to the value of b. If the length of the link is fixed, the effect of increasing the number of links is to decrease b (eqn (3.3a)) and hence to *increase* the breadth of the distribution, and vice versa.

3.5. The distribution of *r*-values

The observation that the maximum probability *per unit volume*, or probability density, as defined by the function (3.3), occurs when $x = y = z = 0$, i.e. when the two ends of the chain are coincident, must not be taken to imply that the *most probable* distance between the ends of a chain is zero. For the calculation of the probability of any given value of the end-to-end distance r the restriction to a particular direction in space implied by the preceding analysis is inappropriate, and it becomes necessary to take account of all directions equally. We therefore consider the probability that the end B of the chain, whose end A is fixed at the origin O, shall be found within a spherical shell of radius r and thickness dr (Fig. 3.6). The appropriate volume element in this case is therefore the volume of the spherical shell, $4\pi r^2\, dr$, and the required probability $P(r)\, dr$ is obtained by multiplying this by the probability density at the radial distance r, as given by (3.5). Hence

$$P(r)\, dr = (b^3/\pi^{\frac{3}{2}}) \exp(-b^2 r^2) \times 4\pi r^2\, dr$$

or

$$P(r)\, dr = (4b^3/\pi^{\frac{1}{2}}) r^2 \exp(-b^2 r^2)\, dr. \tag{3.9}$$

The function $P(r)$, as shown in Fig. 3.5(b), is of an entirely different form from the probability density function (3.5) or the function $p(x)$ representing the probability of a given component of length in a

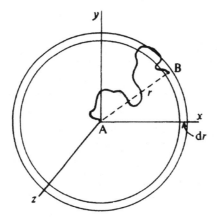

FIG. 3.6. Determination of distribution of r-values, regardless of direction in space.

fixed direction in space. It represents the distribution of end-to-end distance r for a set of free chains such as might exist for example in dilute solution. Whereas the function $p(x)$ is symmetrical and has a maximum value when $x = 0$, the function $P(r)$ is zero at $r = 0$ and reaches a maximum at a finite value of r. This most probable value r_{mp}, found by differentiation of $P(r)$ with respect to r, is

$$r_{mp} = 1/b = (2nl^2/3)^{\frac{1}{2}}. \qquad (3.10)$$

A more important quantity, the *mean-square* value of r, is given by

$$\overline{r^2} = \frac{\displaystyle\int_0^\infty rP(r)\,\mathrm{d}r}{\displaystyle\int_0^\infty P(r)\,\mathrm{d}r} = \frac{3}{2b^2} = nl^2. \qquad (3.11)$$

The functions $P(r)$ and $p(x)$ are identical in form to the corresponding Maxwell distribution functions representing the distribution of velocities among the molecules of a gas. The function $P(r)$ corresponds to the distribution of total velocities v, and the function $p(x)$ to the distribution of the components of velocity v_x with respect to a fixed direction in space. From the mathematical standpoint the two problems are entirely analogous, being concerned with the random addition of vectors representing in the one case a length and in the other a velocity.

Target analogy

The apparent paradox that the probability density function (3.5) is a maximum when $r = 0$, while the distribution function $P(r)$ is zero at $r = 0$ illustrates the necessity for a careful definition of the sense in which the term *probability* is being applied in any given situation. The present problem may be illuminated by considering a strictly analogous problem in everyday life, i.e. the distribution of shots on a target, which for a sufficiently large number of shots is describable by an analogous two-dimensional Gaussian distribution function. The number of shots per unit (small) area is obviously a maximum at the centre ($r = 0$), but if the target area is divided into a number of concentric rings of thickness dr, the maximum number of shots within a ring will occur at a finite value of r, owing to the increase in the area of the ring with increasing r (Fig. 3.7). In this sense it would

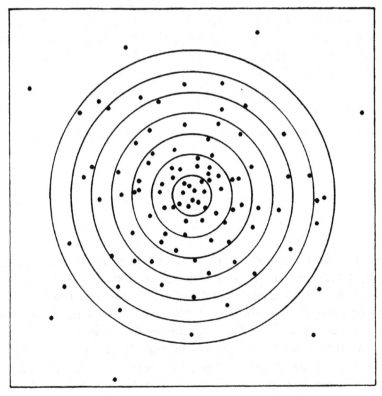

FIG. 3.7. Target analogy to random-chain problem.

be true to say that the most probable position of any shot is at the centre ($r = 0$) but that the most probable distance of any shot from the centre is certainly not zero.

3.6. Equivalent random chain

It may be shown quite generally that the statistical distribution of end-to-end distances for any chain, whatever its geometrical structure, reduces to the Gaussian form if the number of rotatable links is sufficiently large (Flory 1969, p. 6). This is true even when the rotation about single bonds is not completely free but is restricted by 'energy barriers' or other hindrances to rotation. The effect of the structure of the chain on its statistical properties is therefore represented simply by the numerical value of the parameter b in the Gaussian distributions (3.3) or (3.9). Since this parameter also determines the value of the mean-square length through eqn (3.11), it follows that if the mean-square length can be calculated the Gaussian statistical distribution is completely determined.

On the basis of these considerations it can be seen that for any actual long-chain molecule it is always possible to find a corresponding randomly jointed chain which will have the same statistical properties. The first requirement is that the mean-square lengths of the actual molecule and of the postulated random chain shall be the same; this, however, from eqn (3.11), determines only the product nl^2, and not the separate values of n and l. To determine these the further condition, namely, that the two chains—the actual molecule and the postulated randomly jointed chain—shall have the same fully extended or outstretched length, must be introduced. Given these two conditions there is one, and only one, random chain which can be regarded as statistically equivalent to the actual molecule.

To determine the number of links and the length of the link in the equivalent random chain, let us assume that the actual molecule, when fully extended *without distortion of bond lengths or valence angles*, has the length R_m and that its mean-square length is $\overline{r_m^2}$. Let R_r and $\overline{r_r^2}$ be the corresponding values of the fully extended length and mean-square length of the postulated random chain, having n_r links of length l_r. If the mean-square lengths for the two chains are to be the same we have

$$\overline{r_m^2} = \overline{r_r^2} = n_r l_r^2, \tag{3.12}$$

and if both are to have the same fully extended length

$$R_m = R_r = n_r l_r. \tag{3.13}$$

For given values of R_m and $\overline{r_m^2}$ these two simultaneous equations lead to the following solution for n_r and l_r,

$$n_r = R_m^2 / \overline{r_m^2}, \qquad l_r = \overline{r_m^2} / R_m. \tag{3.14}$$

The values of n_r and l_r define a random chain which has the same statistical properties in the Gaussian region and the same fully extended length as the actual molecule, and may therefore be regarded, at least to a first approximation, as statistically equivalent to the actual molecule.

The equivalent random link

The significance of the above concept of the equivalent randomly jointed chain may be illustrated by reference to the idealized models of the polymethylene and *cis*-polyisoprene structures, assuming *free rotation* about bonds. For the former, the mean-square length, from eqns (3.1) and (3.1a) is

$$\overline{r_m^2} = n_m l_m^2 \frac{1 + \cos\theta}{1 - \cos\theta} = 2 n_m l_m^2, \tag{3.15a}$$

while the fully extended length (or length of chain axis for the planar zig-zag form shown in Fig. 3.2(a)) is

$$R_m = n_m l_m \cos\tfrac{1}{2}\theta = \sqrt{\tfrac{2}{3}} n_m l_m. \tag{3.15b}$$

Substitution of these values in eqn (3.14) gives for the equivalent random chain

$$n_r = \tfrac{1}{3} n_m, \qquad l_r = \sqrt{6} l_m = 2\cdot45 l_m. \tag{3.15c}$$

The polymethylene chain (with freely rotating bonds) may therefore be replaced by a randomly jointed chain having one-third the number of links; it follows, therefore, that one random link is equivalent to three C—C bonds in the original structure.

For the *cis*-polyisoprene chain the length of the isoprene unit (in the planar configuration, Fig. 1.2(a) (p. 4) is $0\cdot460$ nm; the fully extended length R_m for a chain containing n_m isoprene units is therefore $0\cdot460\,n_m$. There being four chain bonds per isoprene unit, the corresponding mean-square length in nm^2, from Wall's formula (eqn (3.2)), is $\overline{r_m^2} = 4(0 \cdot 201)\ n_m$ or $0 \cdot 162\ n_m$. These figures give

$$n_r = 1 \cdot 31 n_m, \qquad l_r = 0 \cdot 352 l_m. \tag{3.16}$$

In this case, therefore, the number of monomers per random link is $1/1 \cdot 31$ or $0 \cdot 77$.

It is important to emphasize that these examples refer only to the idealized freely rotating models of the polymethylene and poly(cis-isoprene) chains. The actual properties of the polymethylene and rubber chains, which are examined in later chapters, are significantly different, owing to the presence of energy barriers to rotation, referred to earlier in this chapter.

3.7. The entropy of a single chain

In the case where rotation about the bonds in the molecular chain can be considered to be unrestricted the internal energy of the molecule will be the same for all conformations, and the Helmholtz free energy (p. 28) will be determined solely by the entropy term. According to the general principles of statistical thermodynamics, as developed by Boltzmann, the entropy will be proportional to the *logarithm* of the number of configurations available to the system, i.e. to the logarithm of the number of possible configurations corresponding to any specified state.

If the chain is considered to be isolated and completely free from external constraints, all conformations have the same *a priori* probability. This condition is approximately realized in the case of a molecule dissolved in a large excess of a neutral liquid, i.e. a liquid in which there are no resultant energetic interactions between the polymer and solvent molecules. If, on the other hand, there are certain restrictions imposed on the free motion of the chain, the number of available conformations will be reduced. In a vulcanized rubber, with which we are ultimately concerned, the original molecules are connected together at certain points so as to form a network. The points of cross-linkage thus restrict the motion of the ends of the intervening segments of molecules or 'network chains' to a small element of volume in the immediate vicinity of certain points, which as we shall see later (Chapter 4) may be regarded as fixed in space. This situation can be treated by assuming one end of the chain to be fixed at the origin O (Fig. 3.4) while the other is confined to a small volume element dτ in the neighbourhood of the point P at a distance r from O. In the presence of this restriction, the number of conformations available to the chain is proportional to the probability density (eqn (3.5)) multiplied by the size of the volume element $d\tau$. The entropy s of the chain is therefore given by

$$s = k\{\ln p(x, y, z)\, d\tau\}. \tag{3.17}$$

Substitution of the expression (3.5) for $p(x, y, z)$ thus yields

$$s = k\{\ln (\text{constant}) - b^2 r^2 + \ln d\tau\}. \tag{3.18}$$

Since the volume element $d\tau$ is assumed constant, this may be written in the form

$$s = c - kb^2 r^2, \tag{3.19}$$

where c is an arbitrary constant which includes the volume element $d\tau$. In any actual problem we are concerned only with the difference of entropy between any two states; the constant c is therefore of no physical significance.

The result (3.19) shows the entropy to have its maximum value when the two ends of the chain are coincident ($r = 0$) and to decrease continuously with increasing distance between the ends.

It is important to bear in mind that there is no unique sense in which the entropy of a molecule is to be defined, but that any expression for the entropy (as for the corresponding probability) must be regarded as applicable only in so far as the specified conditions of the problem under examination correspond with the restrictive conditions assumed in the calculation. Entropy is a property of a statistical system, and if the number of links is sufficiently large it is legitimate to regard a single chain as a statistical system. But the chain does not possess an entropy by the mere fact of having a specified conformation (as, for example, in a glassy polymer, where this conformation is frozen in), but only by virtue of the large number of conformations accessible to it under specified conditions of restraint. With this in mind it will be obvious that it would be incorrect to take the r-distribution function (3.9), in which no restriction on the direction of the vector r is involved, as a basis for the calculation of the entropy of a chain whose ends are located at fixed positions in space.

3.8. The tension on a chain

For a chain whose ends are located at specified points separated by a distance r, the entropy s is given by (3.19) and the corresponding Helmholtz free energy by $-Ts$. The work required to move one end of the chain from a distance r to a distance $r + dr$ with respect to the other is equal to the change in Helmholtz free energy (eqn (2.6)), and therefore

$$\frac{dW}{dr} = \frac{dA}{dr} = -T\frac{ds}{dr}. \tag{3.20}$$

Inserting the value of ds/dr obtained by differentiation of the entropy function (3.19) we obtain

$$dW/dr = 2kTb^2r. \tag{3.21}$$

Since work is done in changing the end-to-end distance r, it follows that there must be a tensile force f acting along the direction of r. The work done by this force in a displacement dr being given by $dW = f\,dr$, we have

$$f = dW/dr = 2kTb^2r. \tag{3.22}$$

Thus, a molecule *with its ends fixed at specified points* is acted on by a tensile force in the direction of the line joining its ends and proportional to the length of that line (Fig. 3.8).

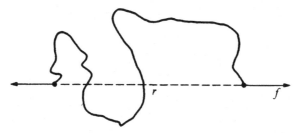

FIG. 3.8. The tension on a chain whose ends are fixed in position is proportional to the distance r.

Since the tension is proportional to r, the molecule may be regarded as possessing an elasticity governed by Hooke's law (stress proportional to strain). It corresponds, in fact, to a classical spring of zero unstrained length.

The spring analogy, however, should not be taken too literally. Being statistical in origin, the tension in a chain whose end-points are located at fixed positions will be subject to continual fluctuations, like the pressure exerted by a gas on the walls of its containing vessel. Likewise, if the molecule is subjected to a constant tension, its length will be a fluctuating quantity. These fluctuations will become relatively less important as the number of links in the chain is increased. The formula (3.22) represents the *average* value of the tension over a period of time.

A further point to be remembered is that the linear force-extension relation (3.22) is subject to the same limitations as the Gaussian distribution function from which it is derived. It is therefore valid only so long as the distance between the ends of the chain is not so large as to be comparable with its fully extended length. This means in practice (as will be shown in Chapter 6) that the distance *r* should not be more than about one-third of the fully extended length. For higher chain extensions the Gaussian approximation becomes increasingly inaccurate and must be replaced by a more accurate distribution function. This is discussed in Chapter 6.

THE ELASTICITY OF A MOLECULAR NETWORK

4.1. The nature of the problem

THE general form of the force–extension curve for a well-vulcanized rubber in simple extension is shown in Fig. 1.1 (p. 2). The abscissae in this diagram represent the extension as a percentage of the unstrained length, and the ordinates the tensile force per unit area of the *unstrained* cross-section. This representation of the force is used for experimental convenience, but the quantity so defined differs from the true stress, or force per unit area measured in the *strained* state. In classical elasticity theory, which is limited to small strains, the difference is unimportant since the area does not change significantly on straining, but in rubbers the difference between the force per unit unstrained area (or nominal stress) and the true stress is considerable.

Whilst most practical tests are concerned with the properties of rubber in simple extension, the stress–strain relations for other types of strain, such as, for example, simple shear or uniaxial compression, are of equal interest from the theoretical standpoint. One of the most interesting aspects of the statistical theory is that it has provided a basis for the correlation of the behaviour of rubbers under different types of strain. In fact, it has led to the derivation of general formulae for the principal stresses in terms of the most general type of strain, from which any particular stress–strain relationship, e.g. for simple extension or shear, is obtainable as a special case. This ability to deal with any type of strain, in contrast to the restriction to simple extension which was characteristic of the earlier theories referred to in Chapter 1, is one of the outstanding merits of the statistical theory and is largely responsible for the great influence which this theory has had on the historical development of the study of large elastic deformations. At the same time, the greater range of opportunities thus presented for comparison with experimental observations has provided a more extensive basis for the critical assessment both of the range of validity of the theory and of its inadequacies and shortcomings.

As explained in earlier chapters, the theory is based on the concept of a vulcanized rubber as an assembly of long-chain molecules linked together at a relatively small number of points so as to form an irregular three-dimensional network. The statistical treatment of the network is similar in principle to the treatment of the single chain given in the last chapter; it is required first to calculate the entropy of the whole assembly of chains as a function of the macroscopic state of strain in the sample, and from this to derive the free energy or work of deformation. From the work of deformation corresponding to a given state of strain the associated stresses are then readily derived by the application of mechanics.

The first successful attack on this problem was due to Kuhn (1936), who derived a relation between the elastic modulus and the molecular weight of the chains, though he did not develop the form of the stress–strain relation for large strains. More precise treatments, leading to explicit forms of stress–strain relations which are valid for large strains, were developed by Wall (1942), Flory and Rehner (1943), James and Guth (1943), and the author (Treloar 1943a, 1943b). These later theories led to results which are substantially in agreement with each other.

4.2. Detailed development of the theory

Fundamental assumptions

The following presentation of the theory is based essentially on the original theory of Kuhn, as amended by the author. It makes use of the following assumptions.

1. The network contains N chains per unit volume, a chain being defined as the segment of molecule between successive points of cross-linkage.
2. The mean-square end-to-end distance for the whole assembly of chains in the unstrained state is the same as for a corresponding set of free chains, and is given by the formula (3.11) (p. 51).
3. There is no change of volume on deformation.
4. The junction points between chains move on deformation as if they were embedded in an elastic continuum. As a result the components of length of each chain change in the same

ratio as the corresponding dimensions of the bulk rubber. (Affine deformation assumption.)

5. The entropy of the network is the sum of the entropies of the individual chains, the latter being given by the formula (3.19) (p. 56).

The following comments on these assumptions may be made here.

Assumption 2 is the most obvious simple assumption that can be made concerning the mean-square value of r in the unstrained state, but it is not necessarily a correct assumption. Its significance will be considered in more detail later.

Assumption 3 is well justified on the basis of experiments, which show the volume changes on deformation to be very small, i.e. of the order 10^{-4} or less.

The affine deformation assumption (4) is the key assumption in the whole theory, since it relates the deformation of the individual chains to the macroscopic strain in the material. Though introduced as an assumption by most authors, its validity has been rigorously proved in the form of the theory given by James and Guth (1943). These authors also show that it is justifiable to neglect the fluctuations of positions of the junction points (arising from the thermal agitation of the associated chains), and to regard them as fixed at their mean or most probable positions.

The last assumption (5) is equivalent to the statement that the chains are 'Gaussian'.

Calculation of entropy of deformation

Most authors have been concerned with the case of a simple extension, but it is no more difficult to treat the more general case of a pure homogeneous strain of any type. This is defined by three principal extension ratios along three mutually perpendicular axes. Under such a strain a unit cube (Fig. 4.1) is transformed into a rectangular parallelepiped having three unequal edge lengths λ_1, λ_2, and λ_3. These extension ratios may be either greater than 1, corresponding to a stretch, or less than 1, corresponding to a compression, provided that the condition for constancy of volume (assumption 3 above), namely,

$$\lambda_1\lambda_2\lambda_3 = 1 \qquad (4.1)$$

is satisfied.

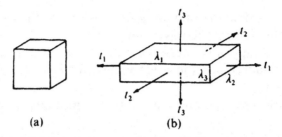

FIG. 4.1. Pure homogeneous strain: (a) the unstrained state; (b) the strained state.

Let us consider an individual chain (Fig. 4.2) having an end-to-end distance represented by the vector r_0, with components (x_0, y_0, z_0), in the unstrained state of the network, and let (x, y, z) be the components of the vector length r of the same chain after deformation. Then by the affine deformation assumption

$$x = \lambda_1 x_0, \qquad y = \lambda_2 y_0, \qquad z = \lambda_3 z_0, \qquad (4.2)$$

the axes of coordinates being chosen to coincide with the principal axes of strain. The entropy of the chain in the original state, as given by eqn (3.19), will be

$$s_0 = c - kb^2 r_0^2 = c - kb^2(x_0^2 + y_0^2 + z_0^2), \qquad (4.3)$$

where b^2 is the constant in the Gaussian distribution function (3.3). The entropy of the same chain in the strained state is obtained by substituting the values of (x, y, z) given by (4.2) for (x_0, y_0, z_0), i.e.

$$s = c - kb^2(\lambda_1^2 x_0^2 + \lambda_2^2 y_0^2 + \lambda_3^2 z_0^2). \qquad (4.3a)$$

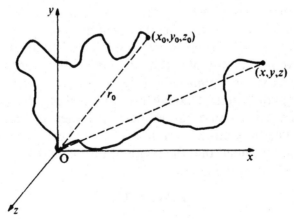

FIG. 4.2. The 'affine' deformation of chains.

The contribution to the total entropy of deformation for the network due to this chain will therefore be

$$\Delta s = s - s_0 = -kb^2\{(\lambda_1^2 - 1)x_0^2 + (\lambda_2^2 - 1)y_0^2 + (\lambda_3^2 - 1)z_0^2\}. \quad (4.4)$$

The total entropy for all the N chains contained in unit volume of the network is obtained by summation of the expression (4.4). Assuming for the moment that the chain *contour length* or chain molecular weight is the same for all the chains, the parameter b (which is a function of chain length) will be constant, and we may write for the total entropy of deformation

$$\Delta S = \sum \Delta s = -kb^2\{(\lambda_1^2 - 1)\sum x_0^2 + (\lambda_2 - 1)\sum y_0 + (\lambda_3^2 - 1)\sum z_0^2\}. \quad (4.5)$$

In this expression $\sum x_0^2$ is the sum of the squares of the x_0-components, in the unstrained state of the network, for the assembly of N chains. Since the directions of the chain vectors r_0 in the unstrained state are entirely random, there will be no preference for the x, y, or z directions and hence, remembering that

$$\sum x_0^2 + \sum y_0^2 + \sum z_0^2 = \sum r_0^2,$$

we may write

$$\sum x_0^2 = \sum y_0^2 = \sum z_0^2 = \tfrac{1}{3}\sum r_0^2. \quad (4.6)$$

But,

$$\sum r_0^2 = N\overline{r_0^2}, \quad (4.6a)$$

where $\overline{r_0^2}$ is the mean-square length of the chains in the unstrained state. Hence from (4.5),

$$\Delta S = -\tfrac{1}{3}Nkb^2\overline{r_0^2}(\lambda_1^2 + \lambda_2^2 + \lambda_3^2 - 3). \quad (4.7)$$

If we now introduce the assumption that the mean-square chain-vector length in the unstrained state is the same as for a corresponding set of free chains we have $\overline{r_0^2} = 3/2b^2$, which on insertion into (4.7) gives

$$\Delta S = -\tfrac{1}{2}Nk(\lambda_1^2 + \lambda_2^2 + \lambda_3^2 - 3). \quad (4.8)$$

This result does not contain the parameter b, and hence is not directly dependent on the chain contour length. It follows that the same formula would apply even if the contour lengths of the chains were not the same for all, as would be the case for a random cross-linking process.

Work of deformation

The Helmholtz free energy or work of deformation is obtainable directly from (4.8). Assuming, in accordance with the basic principles of the kinetic theory, that there is no change of internal energy on deformation (Chapter 2), we have $W = -T \Delta S$ and hence

$$W = \tfrac{1}{2} NkT(\lambda_1^2 + \lambda_2^2 + \lambda_3^2 - 3), \tag{4.9}$$

in which W represents the work of deformation or elastically stored free energy per unit volume of the rubber (later referred to as the *strain-energy function*).

It will be convenient for many purposes to write (4.9) in the form

$$W = \tfrac{1}{2} G(\lambda_1^2 + \lambda_2^2 + \lambda_3^2 - 3), \tag{4.9a}$$

where

$$G = NkT. \tag{4.9b}$$

4.3. Significance of theoretical conclusions

The result represented by eqns (4.9) or (4.9a) is of the utmost importance. This relation may be thought of as the fundamental expression by which the elastic properties of a rubber in the Gaussian region are completely defined, and we shall see in § 4.4 below that it enables the stress–strain relations for any type of strain to be derived. It is of particular interest to observe that this equation (and hence the stress–strain relations derivable from it) involves only a single physical parameter or elastic constant G which is dependent on the structure of the material. This constant is related to N, the number of chains per unit volume, which is itself determined by the degree of cross-linking; for normal cross-linking (in which four chains meet at each junction point) it is equal to twice the number of cross-links per unit volume. It follows that the elastic properties of a rubber, in so far as they can be represented by the statistical theory, are independent of the chemical nature of the molecules of which it is composed, provided only that they have the necessary length and flexibility for the basic assumptions of the theory to be satisfied.

It will be shown later that the elastic constant G in eqns (4.9a) and (4.9b) is equivalent to the shear modulus. Since the length of the network chains decreases as the degree of cross-linking is increased, this constant may be expressed alternatively in terms of the *number average* chain molecular weight M_c. The appropriate relationship (which is easily derived) is

$$G = NkT = \rho RT/M_c, \qquad (4.9c)$$

in which ρ is the density of the rubber and R the gas constant per mole.

4.4. The principal stresses

In the state of pure homogeneous strain depicted in Fig. 4.1, there are three principal stresses acting in directions parallel to the principal axes of strain on planes corresponding to the faces of the rectangular block. If these stresses are denoted by t_1, t_2, and t_3, the problem is to determine their magnitude in terms of the corresponding principal extension ratios λ_1, λ_2, and λ_3. The derivation of these general stress–strain relations is based on the expression (4.9a) for the elastically stored energy, and makes use of the condition for constancy of volume (eqn (4.1)).

As noted earlier, it is important to distinguish between *true* and *nominal* stresses. We shall define t_1, t_2, and t_3 as the forces per unit *strained* area, or true stresses, and use the symbols f_1, f_2, and f_3 to represent the corresponding forces per unit *unstrained* area, or nominal stresses. For a body initially in the form of a unit cube, the force f_1 acts on an area $\lambda_2\lambda_3$ measured in the strained state; the corresponding true stress is therefore, from (4.1),

$$t_1 = f_1/\lambda_2\lambda_3 = \lambda_1 f_1, \qquad (4.10)$$

with similar expressions for t_2 and t_3.

To determine the forces, the method used is to equate the change in the stored energy in any variation of λ_1, λ_2, and λ_3 to the work done by the applied forces. However, these three extension ratios cannot be varied independently of one another, since they are connected by the constant-volume relation (4.1), which implies that if any two of these are chosen as independent variables the third is then determined. Thus if λ_1 and λ_2 are chosen as independent variables we have $\lambda_3 = 1/\lambda_1\lambda_2$. The strain energy function may thus be written

$$W = \tfrac{1}{2}G(\lambda_1^2 + \lambda_2^2 + 1/\lambda_1^2\lambda_2^2). \tag{4.11}$$

Let us consider, for simplicity, the case when only two forces f_1 and f_2 are applied, so that $f_3 = t_3 = 0$ (Fig. 4.3). If the extension in the

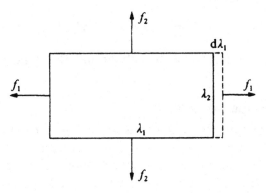

FIG. 4.3. Two-dimensional extension under the action of forces f_1 and f_2.

direction λ_1 is increased by the amount $d\lambda_1$, while λ_2 is held constant, the only work done by the applied forces is that done by the force f_1. Assuming the unstrained block to be in the form of a unit cube we have then

$$dW = f_1\, dl = f_1\, d\lambda_1, \tag{4.12}$$

where dl is the increase in length. From (4.11) we have also

$$dW = (\partial W/\partial\lambda_1)\, d\lambda_1 = G(\lambda_1 - 1/\lambda_1^3\lambda_2^2)\, d\lambda_1. \tag{4.13}$$

Equating these values of dW we obtain

$$f_1 = G(\lambda_1 - 1/\lambda_1^3\lambda_2^2). \tag{4.14}$$

The corresponding principal stress t_1, from (4.10), is

$$t_1 = \lambda_1 f_1 = G(\lambda_1^2 - 1/\lambda_1^2\lambda_2^2) = G(\lambda_1^2 - \lambda_3^2). \tag{4.15}$$

A similar result is obtained for t_2. The three principal stresses thus become

$$t_1 = G(\lambda_1^2 - \lambda_3^2); \qquad t_2 = G(\lambda_2^2 - \lambda_3^2); \qquad t_3 = 0. \tag{4.16}$$

To obtain the general solution corresponding to the case when t_3 is not zero, let us superimpose on to the above stress system a negative hydrostatic pressure $-p$ (i.e. a hydrostatic tensile stress $+p$). Since in accordance with Assumption 3 (§ 4.2) the volume must remain constant this hydrostatic stress can have no effect on the state of strain. The stresses, however, are each increased by the amount p to give

$$t_1 = G(\lambda_1^2 - \lambda_3^2) + p; \qquad t_2 = G(\lambda_2^2 - \lambda_3^2) + p; \qquad t_3 = p. \qquad (4.17)$$

The interpretation of eqns (4.17) is that for a material which is incompressible with respect to volume the stresses are indeterminate to the extent of an arbitrary hydrostatic pressure p. Only the *differences* between any two of the principal stresses, which are of course unaffected by the addition of a hydrostatic stress, are determinate. From either (4.16) or (4.17) these are given by

$$t_1 - t_2 = G(\lambda_1^2 - \lambda_2^2),$$

$$t_2 - t_3 = G(\lambda_2^2 - \lambda_3^2), \qquad (4.18)$$

$$t_3 - t_1 = G(\lambda_3^2 - \lambda_1^2).$$

In most practical problems the fact that the principal stresses are, in general, indeterminate does not lead to any difficulty, since one of them can usually be obtained from the boundary conditions. In particular, if one of the surfaces is stress-free as, for example, in the case of simple extension, the corresponding principal stress is zero, and the indeterminacy is automatically removed. For a simple extension corresponding to an extension ratio λ_1 we have $t_2 = t_3 = 0$, and also, from eqn (4.1), $\lambda_2^2 = \lambda_3^2 = 1/\lambda_1$. Substitution into the first of eqns (4.18) then yields

$$t_1 = G(\lambda_1^2 - 1/\lambda_1). \qquad (4.19)$$

Making use of (4.10) the corresponding force per unit unstrained area becomes

$$f_1 = G(\lambda_1 - 1/\lambda_1^2) \qquad (4.19a)$$

4.5. Significance of single elastic constant

At first sight it seems surprising that a single elastic constant should be sufficient to define the properties of a rubber, in which the strains may be large, whereas in the classical theory of elasticity, which is limited to small strains, two independent elastic constants, e.g. the bulk modulus k and the shear modulus n, are required. This apparent inconsistency is a result of the assumption of constancy of volume or incompressibility. From the physical standpoint this assumption implies that changes of volume, which are proportional to $1/k$, are negligible compared with changes of shape, which are proportional to $1/n$, i.e. that the ratio n/k is small compared with unity. For an ordinary solid n and k are of the same order of magnitude, but for rubbers the ratio of shear modulus to bulk modulus is of the order 10^{-4}, and this condition is therefore satisfied. As a result the response to a stress is effectively determined solely by the shear modulus n. This difference in the values of the two elastic constants is a consequence of the totally different molecular mechanisms responsible for the changes of volume and changes in shape, respectively, to which attention has already been drawn in Chapter 2 (cf. also § 4.7 below).

4.6. The elastic properties of a swollen rubber

The phenomenon of the swelling of rubbers in low molecular weight liquids will be dealt with in Chapter 7. Therefore in this section we shall not consider the process of swelling as such but will restrict ourselves to the consideration of the effect of the swelling on the mechanical properties of a cross-linked rubber. To define the problem it is assumed that the sample in the unswollen state is in the form of a cube of unit edge length, and that it contains N chains (per unit volume). The degree of swelling will be defined in terms of the volume fraction v_2 of rubber in the mixture of rubber and liquid; the volume swelling ratio, referred to the dry state, will thus be $1/v_2$, and the corresponding linear dimensions of the swollen sample (linear swelling ratio) will be $\lambda_0 = 1/v_2^{\frac{1}{3}}$. In the following argument we shall not be concerned with the nature of the swelling liquid, nor with the question of the equilibrium degree of swelling and the factors by which it is determined; the parameter v_2 will be introduced merely as a means of defining the state of swelling of the network, regardless of whether or not this state is the equilibrium state with respect to the absorption of liquid.

The swelling process itself corresponds to an isotropic expansion of the network, and will therefore be accompanied by a reduction in the network entropy. On the application of a stress to the swollen rubber there will be a further reduction of entropy due to the deformation of the already swollen network. The total reduction of entropy involved in the transformation from the initial unswollen unstrained state to the final strained swollen state will thus be the sum of two terms: one associated purely with the swelling, and the other with the subsequent strain. Our present interest will be limited to a consideration of the second of these two terms, this being the one by which the mechanical properties of the swollen rubber are determined.

The expression (4.8) for the entropy of deformation, though originally worked out for the case of a constant-volume deformation, is equally applicable to the more general case in which this restriction is not introduced, since its derivation is not in any way dependent on this limitation. The total entropy change $\Delta S_0'$ in passing from the unstrained unswollen state to the strained swollen state for the most general pure homogeneous strain may therefore be written in the form

$$\Delta S_0' = -\tfrac{1}{2}Nk(l_1^2 + l_2^2 + l_3^2 - 3), \tag{4.20}$$

where l_1, l_2, and l_3 are the lengths of the edges of the original unit cube, i.e. the principal extension ratios referred to the unswollen state. The change of entropy ΔS_0 associated with the initial isotropic swelling, corresponding to the linear extension ratio λ_0, is

$$\Delta S_0 = -\tfrac{1}{2}Nk(3\lambda_0^2 - 3) = -\tfrac{1}{2}Nk(3v_2^{-\frac{2}{3}} - 3). \tag{4.21}$$

The *entropy of deformation* $\Delta S'$ of the swollen network is the difference between these two quantities, i.e.

$$\Delta S' = \Delta S_0' - \Delta S_0 = -\tfrac{1}{2}Nk(l_1^2 + l_2^2 + l_3^2 - 3v_2^{-\frac{2}{3}}). \tag{4.22}$$

We now define extension ratios λ_1, λ_2, and λ_3 with reference to the unstrained *swollen* state, so that $l_1 = \lambda_1\lambda_0 = \lambda_1 v_2^{-\frac{1}{3}}$, etc., and obtain

$$\Delta S' = -\tfrac{1}{2}Nkv_2^{-\frac{2}{3}}(\lambda_1^2 + \lambda_2^2 + \lambda_3^2 - 3). \tag{4.23}$$

This is the entropy of deformation per unit volume of the original unswollen rubber. We require the entropy of deformation ΔS per unit volume of the *swollen* rubber; this is given by

$$\Delta S = v_2 \Delta S' = -\tfrac{1}{2}Nkv_2^{\frac{1}{3}}(\lambda_1^2 + \lambda_2^2 + \lambda_3^2 - 3). \tag{4.24}$$

The corresponding strain-energy function for the swollen rubber thus becomes

$$W = -T\Delta S = \tfrac{1}{2}NkTv_2^{\frac{1}{3}}(\lambda_1^2 + \lambda_2^2 + \lambda_3^2 - 3). \tag{4.25}$$

This equation represents the properties of the swollen rubber in terms of the extension ratios measured in the swollen state and the volume fraction of rubber v_2. It should be noted, however, that N is the number of chains per unit volume of the *unswollen* rubber.

On putting $v_2 = 1$, this result (4.25) reduces to that previously obtained for a dry rubber (eqn (4.9)).

Comparison with dry rubber

Comparison of the result expressed by eqn (4.25) with the corresponding expression for the unswollen rubber shows the dependence of the stored energy on strain to be of the same form in both cases, the only difference being in the value of the shear modulus. If G and G' are the respective moduli in the unswollen and in the swollen states, then

$$G' = G v_2^{\frac{1}{3}} = \frac{\rho RT}{M_c} v_2^{\frac{1}{3}}, \qquad (4.26)$$

where ρ is the density in the unswollen state. The resultant stress–strain relations for the swollen rubber, which are of the type

$$t_1 - t_2 = G v_2^{\frac{1}{3}}(\lambda_1^2 - \lambda_2^2) = G'(\lambda_1^2 - \lambda_2^2), \qquad (4.27)$$

show a corresponding difference. These results imply that the only effect of the swelling is to reduce the modulus in inverse proportion to the cube root of the swelling ratio, without changing the form of the stress–strain relations.

Alternative formula for network entropy

By the use of a different method of calculating the entropy of deformation, based on the consideration of the entropy of formation of the network from a corresponding set of independent chains, Flory (1950) and Wall and Flory (1951) have obtained the result

$$\Delta S_0' = -\tfrac{1}{2}Nk\{l_1^2 + l_2^2 + l_3^2 - 3 - \ln(l_1 l_2 l_3)\}, \qquad (4.28)$$

which differs from (4.20) by the inclusion of the term $-\ln(l_1 l_2 l_3)$. The method of derivation of this formula has been the subject of much discussion, which, however, remains inconclusive. Fortunately the two formulae (4.20) and (4.28) become identical for the case of an unswollen rubber, for which $\ln(l_1 l_2 l_3) = 0$, while for a swollen rubber the effect of the additional term is in most practical cases rather small.

4.7. Development of the theory by James and Guth

In this section we shall examine more critically the assumptions on which the network theory as presented above was based, and consider some of the refinements which have been suggested with a view to bringing the theoretical model into closer correspondence with physical reality.

The first and most obvious objection to the theory in the form outlined above is that the network junction points are conceived of as fixed at particular points in space. The freedom of movement of the system is thus limited to the lengths of chain between these fixed points. The function used to represent the entropy of the individual chains corresponds precisely with this conception, since it is derived on the basis that the ends of the chain are held at fixed positions (p. 55). In reality, of course, the junction points are not fixed but take part in the micro-Brownian motion of the associated chain elements or links. In the very detailed and precise treatment given by James and Guth (1943) it is only those junction points which are located on the boundary surfaces of the rubber which are specified as fixed; all the other junction points are allowed complete statistical freedom, subject only to the restraints imposed by the associated network chains through their mutual interconnections. The system thus comprises a certain number of fixed or boundary junction points, whose relative positions define the state of strain, together with a very much larger number of fluctuating junction points. The statistical problem is to compute the number of configurations, or the configurational entropy, of the whole assembly of chains for specified positions of the boundary junction points.

The solution of this problem involves a rather lengthy mathematical argument, which will not be reproduced. The detailed analysis, however, reveals a number of particularly significant properties of the Gaussian network, of which the most important are the following.

1. The fluctuations of position of any junction point in a network of Gaussian chains may be described by a Gaussian probability function. The mean value of the fluctuations of any given junction point is independent of the strain in the network.

2. The average force between any two adjacent junction points is the same as if both were fixed at their most probable positions.
3. The average forces exerted by the network are the same as would be produced if each chain were replaced by a classical elastic spring exerting a tension proportional to its length.
4. As a corollary to point 3, if the network is subjected to a homogeneous strain, the average positions of the junction points will be displaced as if they were embedded in an elastic continuum.
5. The forces exerted by the network are the same whether any given junction point is treated as free, or as fixed at its most probable position.
6. For the calculation of the forces acting on the bounding surfaces, the network may be replaced by a fictitious system of three independent sets of Gaussian chains parallel to the three coordinate axes.

These and other results are proved by a rigorous analysis which is quite general in that it involves no assumptions regarding the distribution of chain contour lengths or the detailed structure of the network. Proposition 4 establishes the validity of the affine deformation assumption used in the elementary presentation of the theory. Proposition 1 implies that the portion of the configurational entropy which is associated with the fluctuations of the junction points is independent of strain, and hence may be ignored in calculating the entropy of deformation (the same principle is involved in proposition 5). This justifies the elementary treatment in which the junction points are considered to be fixed.

In the previous chapter it was shown that a single chain could be compared in its elastic properties with a classical elastic spring, in which the force is proportional to its end-to-end distance. Proposition 3 carries the analogy still further, and implies that each chain in the network is under a tension which (in the absence of any other forces) would tend to reduce the volume of the network to zero. In an actual rubber this contractile tendency is counterbalanced by mutual repulsive forces between the atoms. Thus James and Guth distinguish, in the actual rubber, two quite different types of forces, the first being those associated with the configurational entropy of the network, while the second, arising from van der Waals' interactions between the molecules, determines the volume of the system, or its response to hydrostatic pressure. The assumption involved in the statistical treatment is that the free energy of deformation is associated only with the network configurations, and not at all with the interatomic forces. This is true as a first approximation; so long, in fact, as changes of volume are of negligible amount, so that the rubber may be considered incompressible. When this condition is satisfied an arbitrary hydrostatic pressure may be introduced as required to balance the internal tension due to the network, the value of this pressure being determined by the boundary conditions.

It is seen, therefore, that the more precise analysis of James and Guth justifies most of the assumptions on which the elementary treatment given in § 4.2 was based. There remains, however, one important exception to this general conclusion, which must now be examined more critically. The exception is the assumption that the mean-square end-to-end distance for the chains in the unstrained network is the same as for a corresponding set of free chains. This assumption is related to the rather more specific assumption (which was implied in the early theories of Kuhn (1936) and of Wall (1942), and reproduced in the first edition of this book) that the *distribution* of chain vector lengths in the unstrained network is identical to the instantaneous distribution of chain vector lengths for a corresponding set of free chains. At first sight this appears to be a plausible assumption, for in any assembly of chains, prior to cross-linking, not only the whole molecules, but also all segments within the molecules (provided they contain a sufficient number of links) will be describable in terms of the Gaussian distribution function. If at any instant a number of cross-linkages are simultaneously introduced so as to form a network, the vector-length distribution immediately after cross-linking might be expected to remain unchanged.

This argument may be challenged on two grounds. First, even though the distribution of chain vector lengths immediately after cross-linking were consistent with the above assumption it would not remain so. For the introduction of additional restraints in the form of cross-linkages creates a totally new situation in which the junction points will tend to move to new positions in which they are in equilibrium under the action of the effective tensions in the associated chains—tensions which were non-existent prior to cross-linking. Secondly, in any actual cross-linking reaction the junction points are introduced not simultaneously but progressively, so that the system continuously approaches a new equilibrium state after the introduction of each successive junction point. Under these conditions the original assumption is even less justifiable.

James and Guth (1947) show that the general expression for the modulus may be written in the form

$$G = kT \sum_\tau n_\tau \lambda_\tau^2, \tag{4.29}$$

where n_τ is the number of links in the τth chain, and λ_τ is its mean fractional extension r/nl in the unstrained state of the network. In the case when the distribution of vector lengths in the unstrained state is equal to that for a corresponding set of free chains this expression reduces to the form

$$G = NkT \tag{4.29a}$$

in agreement with the elementary theory. An attempt to evaluate the expression (4.29) without introducing this assumption is discussed in § 4.9 below.

4.8. Network imperfections: 'loose end' corrections

In the preceding treatment of the network it has been tacitly assumed that the molecules originally present in the rubber are cross-linked in such a way as to form an assembly of mutually interconnected segments or network 'chains'. For normal 'tetrafunctional' cross-linking, in which four network chains terminate at each junction point, the number of chains will be exactly twice the number of junction points, and every chain will contribute equally to the network elasticity. This is obviously an oversimplification, for it is clear that in any actual process involving random cross-linking various types of departure from this idealized structure must be present. Three of the principal types of departure, or 'network defects', to which Flory (1944) has drawn attention, are illustrated in Fig. 4.4. The first (Fig. 4.4(a)) consists of an interlooping or physical entanglement between chains which, by restricting the number of available configurations, has an effect comparable to that of a chemical cross-linkage, and will tend to increase the modulus. The second type of defect (Fig. 4.4(b)) occurs as a result of the linkage of two points on a single chain, giving rise to a closed loop which makes no contribution to the network elasticity; such

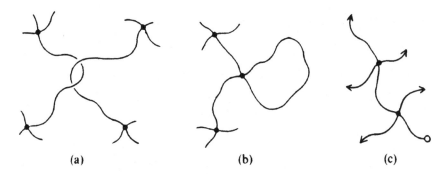

(a)   (b)   (c)

FIG. 4.4. Types of network defect. (Flory 1944.) ●Cross-linkages, ○Chain termination.

intramolecular linkages should therefore be discounted in the evaluation of the modulus. The third type (Fig. 4.4(c)) consists of chains which are connected to the network at one end only; such terminal chains or 'loose ends' likewise make no contribution to the network elasticity.

Of these network defects, only the last was treated quantitatively by Flory. Taking the number of original or primary molecules before cross-linking to be N_p, he argued that $N_p - 1$ intermolecular linkages are required to link these together into a single ramified structure in which there are no closed loops. Thereafter, each additional corss-link will produce one closed loop, or two network chains. It is only these additional cross-linkages which are effective in network formation. Taking ν_0 to be the total number of cross-linkages introduced, the number ν_e of these which are *effective* will therefore be

$$\nu_e = \nu_0 - N_p,$$

where N_p is taken as equivalent to $N_p - 1$. The number N_e of effective chains will be twice this quantity, i.e.

$$N_e = 2\nu_e = 2\nu_0(1 - N_p/\nu_0). \tag{4.30}$$

If no defects were present, we should have $N_e = 2\nu_0$. The additional factor $1 - N_p/\nu_0$ thus represents the effect of the loose ends. It is convenient to express this in terms of the molecular weight M of the primary molecules. For unit volume of rubber of density ρ, we have $N_p = \rho A_0/M$, where A_0 is Avogadro's number, and similarly $2\nu_0 = \rho A_0/M_c$, where M_c is the 'chain molecular weight' for the *idealized* network, in which loose ends are absent. Hence

$$N_e = 2\nu_0(1 - 2M_c/M).$$ (4.31)

The corresponding shear modulus is given by

$$G = N_e kT = \frac{\rho RT}{M_c}\left(1 - \frac{2M_c}{M}\right).$$ (4.32)

It will be seen that this reduces to the elementary form (4.9c) for sufficiently large values of the primary molecular weight, i.e. as $1/M \to 0$.

Later modifications

While Flory's correction for loose ends, represented by eqn (4.32), has been frequently applied in the interpretation of experimental data (cf. Chapter 8), later workers have introduced further modifications based on more realistic treatments of the process of

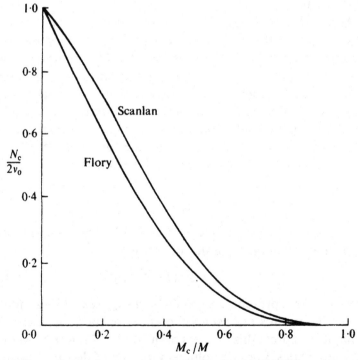

FIG. 4.5. Correction for 'loose ends' as derived by Scanlan (1960) for primary chains of uniform length, compared with Flory's formula. The quantity $N_e/2\nu_0$ represents the fraction of effective cross-linkages.

network formation. These treatments take into account such factors as the distinction between different types of junction points (tri-functional, tetra-functional, etc.), and the effect of a 'sol' component not attached to the coherent network structure. In general, these treatments do not lead to explicit algebraic solutions, but require evaluation by numerical methods. Particular mention may be made of the theories of Case (1960), Scanlan (1960), and Gordon, Kucharik, and Ward (1970). These agree in finding that Flory's formula overestimates the effect of loose ends. The difference depends to some extent on the distribution of chain lengths among the primary molecules, but the results of Scanlan, corresponding to a uniform chain length, may be taken as representative (Fig. 4.5). For high degrees of cross-linking the reduction in modulus due to loose ends is only half that given by Flory's formula (corrected to take account of the presence of a 'sol' component, representing chains not connected to the network). Since for practical vulcanized rubbers the values of M_c/M are typically in the neighbourhood of 0·05, this will apply also to most practical cases. This particular conclusion, moreover, is unaffected by the replacement of the uniform initial chain length by a random distribution, and may therefore be regarded as quite general.

4.9. The absolute value of the modulus

The discounting of ineffective chains which contribute nothing to the network elasticity still leaves open the question of the value to be assigned to the mean-square chain vector length in the unstrained network. This question has been examined in detail by James and Guth (1947), who consider the cross-linking to take place not instantaneously but progressively, the network being allowed to come to equilibrium after the introduction of each successive cross-linkage. They note that the detailed solution to this problem requires a knowledge not only of the number of effective cross-linkages in the final network (in the sense used in the preceding section) but also of the number B_a which were effective or 'active' at the time when they were introduced. This quantity B_a is not obtainable from a knowledge of the final network structure but is a 'historical' parameter.

James and Guth's treatment involves the calculation of the increase in modulus dG/dB_a resulting from the introduction of one cross-link between any two arbitrary points (or segments) of the partially formed coherent network existing at any instant. The result, which is only approximate, is expressed by the relation

$$dG/dB_a \simeq kT. \qquad (4.33)$$

This relation was shown to be exact for two extreme cases, corresponding to (a) very loose and (b) very tight 'coupling' or topological connection between the two points to be cross-linked. It was also shown to be a very close approximation (i.e. to within 4 per cent) for an idealized regular cubic network. It was therefore considered to be substantially correct for the general case.

The result (4.33) may be compared with that given by the more elementary theory, as represented by eqn (4.29a). Since the number of chains is in this case taken to be equal to twice the number of cross-links, eqn (4.29a) would yield an increase of modulus of $2kT$ per cross-link introduced. Eqn (4.33) therefore represents a reduction of one-half compared with the standard formula. This reduction expresses the effect of the subsequent adjustment of the junction-point positions after the introduction of an additional cross-linkage—a process which is necessarily accompanied by an increase of entropy of the whole assembly, and therefore by a reduction in the mean-square vector extension of the chains.

The second stage of James and Guth's treatment consisted in relating the number N_e of effective chains† in the final network to the number B_a of cross-linkages which were 'active' at the time of their formation. Unfortunately this is exceedingly difficult to estimate, since each new cross-link introduced may increase the number of effective chains by 0, 1, 2, *or more*. James and Guth consider that it is 'reasonable to expect' that

$$N_e \simeq 2B_a, \qquad (4.34)$$

this being the relation which would apply at relatively high degrees of cross-linking, if the effects of loose ends were disregarded. Taken together with (4.35) this leads to the value of modulus

$$G \simeq \tfrac{1}{2}N_e kT, \qquad (4.35)$$

which again represents a reduction of one-half compared with the standard formula (4.29a).

---

† Given by G_a in James and Guth's notation.

Criticism of James and Guth's conclusion

While James and Guth's criticism of the elementary theory must be accepted, it is rather difficult to assess the reliability of their final conclusion. Although the result represented by (4.33) is approximate only, the analysis indicates that it should be a fairly close approximation. It is in connection with (4.34) that the more serious uncertainty arises, since this estimate has no rigorous foundation. Bearing this in mind, the question of the precise significance of the factor $\frac{1}{2}$ in (4.35) must remain open to doubt.

Taking all aspects of the problem into account there necessarily remains some uncertainty in the quantitative evaluation of the modulus, since any of the proposed representations of the network must depart to some extent from the conditions corresponding to even the simplest real material. An uncertainty corresponding to a factor of two between different forms of the theory probably represents the best that can be expected in the circumstances, and is in any case not likely to be of great practical importance.

5

EXPERIMENTAL EXAMINATION
OF THE STATISTICAL THEORY

5.1. Introduction

IN the last chapter we saw that the statistical network theory leads to an expression for the work of deformation in a pure homogeneous strain of the most general type (eqn (4.9)), and to corresponding general stress–strain relations (eqns (4.18)), which involve only a single material constant or modulus, which is related to the degree of cross-linking or number of chains per unit volume of the network (eqn 4.9b)). Attempts to verify the theory by comparison with the actual properties of rubbers have been concerned both with the form of the stress–strain relations, and with the comparison of the measured value of the modulus with that derived from chemical estimates of the degree of cross-linking. Only the first of these two aspects—the form of the stress–strain relations—will be considered in the present chapter, the question of the absolute value of the modulus being the subject of Chapter 8.

Reference has already been made to the fact that most of the early studies, being restricted to the case of simple extension, were not capable of providing a critical assessment of the statistical theory, whose particular merit lies in its ability to predict the behaviour to be expected in any type of strain. The main emphasis, therefore, will be on the application of the theory to *different* types of strain. For this purpose, we shall examine certain particularly simple types of strain which are most readily produced and which display the significant features of the phenomena encountered in the most tangible form. These are (1) simple extension, (2) uniaxial compression or equi-biaxial extension, and (3) shear. The more complex problem of the pure homogeneous strain will not be examined at this stage, but will be deferred to Chapter 10, in which the problem of the mechanics of rubber is considered in a more general way.

The theoretical relations corresponding to the above particular types of strain will first be derived.

5.2. Particular stress–strain relations

Simple extension

In this type of strain one dimension of the specimen, which for simplicity may be taken to be a cube of unit edge length (Fig. 5.1), is

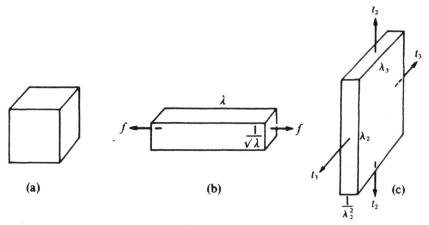

FIG. 5.1. Types of strain. (a) Unstrained state; (b) simple extension; (c) uniform two-dimensional (equi-biaxial) extension.

increased in the ratio λ while the other two dimensions are correspondingly reduced. For constancy of volume the incompressibility condition (eqn (4.1)) then gives

$$\lambda_1 = \lambda; \qquad \lambda_2 = \lambda_3 = \lambda^{-\frac{1}{2}}. \qquad (5.1)$$

The corresponding strain energy, from (4.9a), is therefore

$$W = \tfrac{1}{2}G(\lambda^2 + 2/\lambda - 3). \qquad (5.2)$$

The only force acting is the tensile force in the direction of the extension. If f is the magnitude of this force per unit cross-sectional area *measured in the unstrained state*, the work done in increasing the length by an amount dl is $dW = f \, dl = f \, d\lambda$, hence

$$f = dW/d\lambda = G(\lambda - 1/\lambda^2). \qquad (5.3)$$

The quantity f is the so-called *nominal* stress. Since, from (5.1), the cross-sectional area is reduced in the ratio $1/\lambda$ on straining, the true stress t (i.e. the force per unit area measured in the strained state) is given by

$$t = \frac{f}{1/\lambda} = \lambda f = G(\lambda^2 - 1/\lambda), \tag{5.4}$$

which is the form derived earlier (eqn (4.19)) from the general stress–strain relations.

Uniaxial compression

The state of strain known as uniaxial compression is obtained by the application of inwardly directed forces to a pair of opposite surfaces of a cubical block (or, in the case of a cylinder, in a direction parallel to the axis), the lateral surfaces being free of stress. This type of strain is formally identical to simple extension, but in this case λ (the compression ratio) is less than 1. The expansion of the lateral dimensions is in the ratio $\lambda^{-\frac{1}{2}}$, in accordance with (5.1). The force per unit unstrained area is again given by (5.3), but in this case

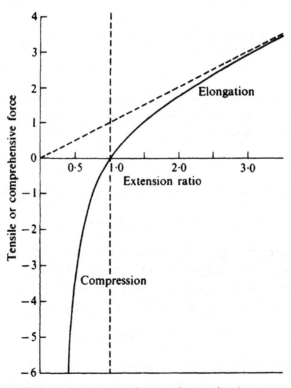

FIG. 5.2. Theoretical relation between force and extension (or compression) ratio (eqn (5.3), with $G = 1 \cdot 0$).

f will be negative (i.e. compressive). Thus both elongation and compression are represented by a single curve, having the form shown in Fig. 5.2.

Equi-biaxial extension (inflation)

In this type of strain the rubber is stretched by equal amounts in two directions at right angles (Fig. 5.1(c)). Such a strain occurs, for example, in the inflation of a spherical balloon. The state of strain corresponds exactly with that produced by uniaxial compression, the only difference being in the nature of the applied stress. If λ_2 is the stretch ratio in the plane of the sheet, we have (as for uniaxial extension or compression, eqns (5.1))

$$\lambda_2 = \lambda_3 = \lambda_1^{-\frac{1}{2}}, \tag{5.5}$$

where λ_1 corresponds to the thickness direction.

Since the stress t_1 normal to the plane of the sheet is zero, the stress t_2 (or t_3) in the plane of the sheet is obtainable from eqns (4.18). Taking into account the relation (5.5) we have, therefore,

$$t_2 = t_3 = G(\lambda_2^2 - \lambda_1^2) = G(\lambda_2^2 - 1/\lambda_2^4). \tag{5.6}$$

This equation represents the true stress (force per unit strained area) in the sheet, as a function of the extension ratio in the two-dimensional extension. In the case of a thin sheet (e.g. as in an inflated balloon) it is of interest to calculate the force f acting on a section of unit length (in the strained state) cut at right angles to the plane of the sheet. If d_0 is the original thickness, the area on which the force f acts is $\lambda_1 d_0$, hence

$$f = t_2 \lambda_1 d_0 = t_2 d_0 \lambda_2^{-2} = G d_0(1 - 1/\lambda_2^6). \tag{5.7}$$

In the equi-biaxial extension all lines in the plane of the extension are changed in the same ratio, and the stress is the same on all sections normal to this plane. For a *thin* sheet this stress may be thought of as a surface tension, or force f per *unit length*, whose magnitude is given by (5.7). For extensions exceeding $\lambda = 2 \cdot 0$ this tension becomes substantially independent of the extension and is therefore analogous to the surface tension of a liquid (Fig. 5.5).

Simple shear

Simple shear is a type of strain which may be represented by the sliding of planes which are parallel to a given plane through a

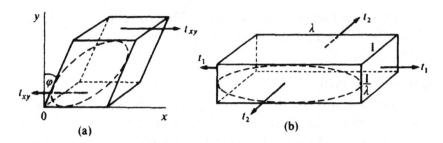

FIG. 5.3. (a) Simple shear; (b) pure shear.

distance proportional to their distance from the given plane (Fig. 5.3(a)). It is *by definition* a constant-volume deformation, whether or not the material is incompressible. The lateral faces of a cube are transformed by simple shear into parallelograms, and the amount of the shear is measured by the tangent of the angle ϕ through which a vertical edge is tilted. There is no strain in the plane normal to the plane $x0y$ (the plane of shear); the extension ratio corresponding to this direction is therefore unity. Since the volume is constant, the three principal extension ratios may therefore be expressed in terms of the major axis of strain λ_1, i.e.

$$\lambda_1 = \lambda_1, \qquad \lambda_2 = 1, \qquad \lambda_3 = 1/\lambda_1. \qquad (5.8)$$

It is to be noted that the directions of the principal axes of strain are not related to the direction of sliding in any simple way, and depend on the magnitude of the strain. (For *small* strains, the principal axes are inclined at 45° to the direction of sliding.) The amount of the shear γ may be related either to the angle ϕ or to the principal extension ratios (Love 1934) thus

$$\gamma = \tan \phi = \lambda_1 - 1/\lambda_1. \qquad (5.9)$$

The strain energy W, from (5.1) and (5.8), has the form

$$W = \tfrac{1}{2}G(\lambda_1^2 + 1/\lambda_1^2 - 2) = \tfrac{1}{2}G\gamma^2. \qquad (5.10)$$

Assuming that the work done on the body is due entirely to the shear stress t_{xy}, it follows that

$$t_{xy} = \mathrm{d}W/\mathrm{d}\gamma = G\gamma, \qquad (5.11)$$

which means that the shearing stress is proportional to the shear strain. Hooke's law is therefore obeyed in simple shear, and the quantity G corresponds to the modulus of rigidity in the classical theory of elasticity. The statistical theory thus leads to the interesting result that a rubber should obey Hooke's law in shear, though not in extension or compression.

Pure shear

For some purposes the less familiar type of deformation known as *pure shear* is more convenient than simple shear. In general, a *pure* strain is one which involves extensions in three perpendicular directions, without rotation of the principal axes of strain; the most general homogeneous strain involves both extensions and rotations. In a pure shear the extension ratios are represented as before by eqns (5.8), but the axes of strain are not rotated (Fig. 5.3(b)). A *simple* shear is thus equivalent to a *pure* shear together with a rotation.

The state corresponding to pure shear, *in the case of an incompressible material*, may be maintained by principal stresses t_1 and t_2 only, with $t_3 = 0$. For a material obeying the relations (4.18) (p. 67) these have the values

$$t_1 = G(\lambda_1^2 - 1/\lambda_1^2), \qquad t_2 = G(1 - 1/\lambda_1^2). \tag{5.12}$$

These are the stresses referred to the strained surface areas. The corresponding forces per unit unstrained area are, from eqn (4.10),

$$f_1 = G(\lambda_1 - 1/\lambda_1^3), \qquad f_2 = G(1 - 1/\lambda_1^2). \tag{5.13}$$

5.3. Experimental examination of stress–strain relations

In examining the applicability of the stress–strain relations deduced above, it is desirable to work with samples of identical material under all types of strain, so that numerical values of the constant G derived from experiments involving different types of strain may be directly compared. Systematic experiments of this kind were originally carried out by the author (Treloar 1944a) on a natural rubber vulcanizate containing 8 per cent of sulphur, this being chosen for its relative freedom from hysteresis compared with more typical commercial vulcanizates available at that time. The types of strain studied were (a) simple extension, (b) equi-biaxial extension, and (c) pure shear. By working with equi-biaxial (two-dimensional) extension rather than the equivalent uniaxial compression the difficulties associated with the bulging of the sample under compressive loading were avoided, and hence very much larger values of equivalent compressive strain could be achieved. Similarly, from experiments on pure shear it was possible to obtain much larger values of strain than would have been practicable for the equivalent simple shear. Furthermore, each of these three types of strain could be carried out on samples cut from a single sheet of vulcanized rubber, so that any variations in modulus which might have arisen from the use of independently vulcanized samples were automatically eliminated.

Simple extension

Fig. 5.4 shows the behaviour in simple extension, obtained by direct loading. Up to about 450 per cent extension ($\lambda = 5.5$) the curves were substantially reversible (curve (c)), but at higher elongations hysteresis effects became apparent (curve (b)). The curve (a), for which the experimental points are shown, was continued up to the breaking point. The theoretical curve (eqn (5.3)) has been fitted to the experimental data in the region of small extensions, the required value of the constant G being $0.39\,\mathrm{N\cdot mm^{-2}}$.

Comparison of the theoretical and experimental curves shows a rather poor agreement. The deviations are of two distinct kinds. First, in the middle region of extension (from $\lambda \simeq 1.5$ to $\lambda \simeq 4$), the experimental curve falls *below* the theoretical one, due to the fact that its curvature is initially greater than it should be according to the theory. Secondly, at high extensions ($\lambda > 4$), the slope of the experimental curve begins to rise, the rate of this rise increasing progressively as the breaking point is approached.

The nature of these deviations, which are more apparent in the case of simple extension than in other types of strain, will be examined in some detail in § 5.4. It will be sufficient here to note that the more prominent of the above two effects, namely, the upturn of the curve in the high-extension region, is not to be regarded as a failure of the statistical theory as such but is the result of a mathematical simplification inherent in the Gaussian statistical theory, based on the distribution function (3.3) (p. 47). As pointed out in Chapter 4, this theory is valid only so long as the extension is not too great; for very large extensions the effect of the finite extensibility of the chains (and hence of the network) must be taken into account. The non-Gaussian theory discussed in the following chapter, in which this effect is taken into account, will be seen to

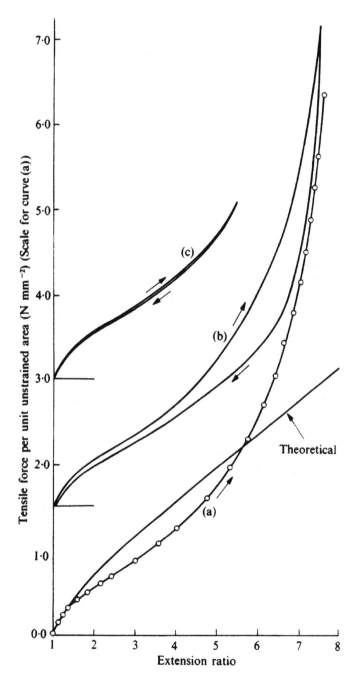

FIG. 5.4. Simple extension. Comparison of experimental curve with theoretical form (eqn (5.3)) ($G = 0.39 \, \text{N} \, \text{mm}^{-2}$).

provide a basis for the interpretation of the upward curvature at large strains.

The remaining deviation—the lower stress in the intermediate region of strain—presents a more serious difficulty, for which no satisfactory explanation has yet been proposed.

Equi-biaxial extension (inflation)

This type of strain was produced by the inflation of a circular rubber sheet clamped round its circumference, after the manner of a bursting test. The strain in the sheet is of course not uniform over its surface, but it is very nearly uniform over a region in the neighbourhood of the centre of the sheet, or 'pole' of the spheroidal balloon, to which the measurements of strain were confined. The measurement of the extension ratio λ_2 in the plane of the sheet was obtained by observing the length of the circular arc between two marked points on the sheet symmetrically disposed with respect to the centre. In addition the radius r of curvature in the polar region and the corresponding inflation pressure p were observed. By analogy with surface tension, the latter is related to the tensile force f per *unit length* of section by the equation

$$p = 2f/r. \qquad (5.14)$$

The result is shown in Fig. 5.5, in which the ordinate is f/d_0, where d_0 is the orginal thickness of the sheet. The continuous curve corresponds to the theoretical relation (5.7). The agreement between theory and experiment is very close up to $\lambda_2 \simeq 3\cdot0$, after which the effects of the limited extensibility of the network begin to appear. It is satisfactory to note that the value of the parameter G which most nearly represents these results is the same as that chosen to fit the first part of the curve for simple extension.

Equivalent uniaxial compression

The experimental values of the force f in the plane of the sheet may be used to calculate the equivalent compressive stress in the corresponding uniaxial compression. According to eqn (5.7) the true stress t_2 in the plane of the inflated sheet is given by

$$t_2 = \lambda_2^2 f/d_0. \qquad (5.15)$$

Superposition of a hydrostatic pressure $-t_2$ reduces the tensile stresses t_2 and t_3 to zero and gives a resultant compressive stress t_1

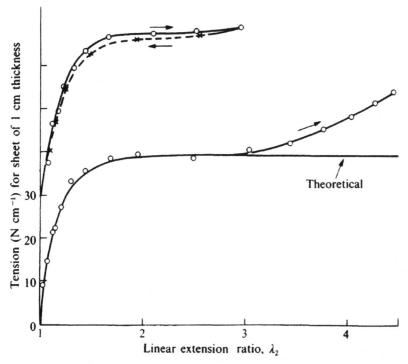

FIG. 5.5. Two-dimensional (equi-biaxial) extension. Comparison of experimental curve with theoretical form (eqn (5.7)) ($G = 0.39\,\mathrm{N\,mm^{-2}}$).

normal to the surface of the sheet of the same magnitude. Hence,

$$t_1 = \lambda_2^2 f / d_0. \qquad (5.16)$$

This is the *true* compressive stress. The corresponding compressive force f_1 acting on unit area measured in the unstrained state is readily obtained. Since unit unstrained area in the plane of the sheet becomes an area λ_2^2 in the strained state, we have

$$f_1 = t_1 \lambda_2^2 = \lambda_2^4 f / d_0. \qquad (5.17)$$

Eqn (5.17) may be used to convert values of f, the tensile force in the plane of the sheet, to f_1, the equivalent compressive force. This may then be plotted against λ_1, the equivalent compression ratio (as given by eqn (5.5)). The result so obtained is shown in Fig. 5.6, together with the first part of the simple extension curve already given in Fig. 5.4. Both the extension and the compression data are seen to lie approximately on a single curve, corresponding to the

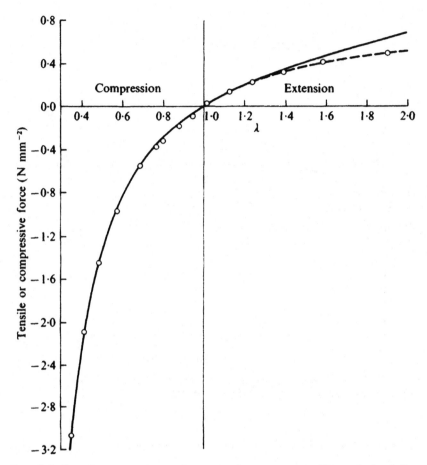

FIG. 5.6. Complete extension and compression curve. ——— Theoretical relation (eqn (5.3)). Compression data from equivalent two-dimensional extension (Fig. 5.5).

theoretical relation (5.3), with no discontinuity in passing through the origin.

It is noteworthy, however, that the compression data fit the theoretical curve over a much greater fraction of the range covered than the data for simple extension.

Pure shear

The definitions of pure shear and simple shear, and the relation between these two types of strain, have been discussed in § 5.2 above. For an incompressible material such as rubber, a state of pure shear may be achieved by the stretching of a rectangular sheet in one direction so as to produce an extension ratio λ_1, while maintaining the perpendicular or transverse dimension unchanged ($\lambda_2 = 1$). Ideally, this necessitates the application of tensile forces f_1 in the direction of the extension and f_2 in the transverse direction. In practice, however, if the width of the sheet is very much greater than its 'length', the direct application of forces in the transverse direction can be dispensed with, these forces being automatically generated as a result of the restraints introduced by the clamps. This is illustrated in Fig. 5.7, which represents the appearance of a wide

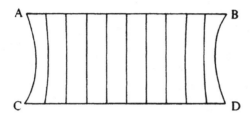

FIG. 5.7. Distribution of strain on stretching of wide sheet. The edges AB and CD are clamped.

strip ($75 \text{ mm} \times 5 \text{ mm}$), on whose surface a set of equidistant vertical lines had been marked, when stretched to about 6 times its original length. It is seen that except in the immediate vicinity of the free edge the state of strain is substantially uniform, corresponding to $\lambda_2 = 1$. (In principle it is possible, by working with strips of equal length but different widths, to eliminate the edge effect; in practice, however, this was found to be an unnecessary complication.)

Measurement of the tensile force f_1 (per unit unstrained area) as a function of the principal extension ratio λ_1, for such a sheet, by the same method as that used in the simple extension experiment, resulted in the curve shown in Fig. 5.8. The theoretical form (5.13), with $G = 0.39 \text{ N mm}^{-2}$, is shown for comparison. It is seen that the deviations from the theoretical form are of a similar kind to those observed in simple extension, though the amount of the difference in the region $\lambda_1 = 1.5{-}4.0$ is relatively somewhat smaller.

Equivalent simple shear

From the measured force f_1 in the principal direction λ_1 for pure shear it is possible to calculate the shear stress in the equivalent simple shear. This could be done by direct resolution of the force,

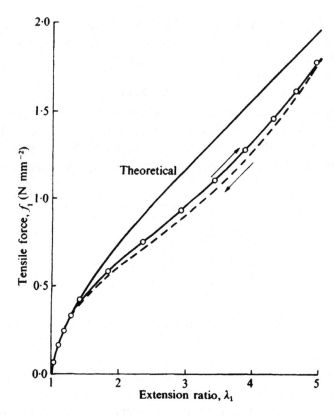

FIG. 5.8. Force–extension relation for wide sheet (pure shear) compared with theoretical relation $f_1 = G(\lambda_1 - 1/\lambda_1^3)$ $(G = 0.39\,\text{N mm}^{-2})$.

but the algebraic transformations are somewhat involved (cf. Chapter 12). A very simple solution may, however, be obtained by making use of the strain energy or work of deformation W, per unit volume. In the case of pure shear the only work done is that done by the force f_1 (since λ_2 is constant). Hence, for an incremental extension $d\lambda_1$,

$$dW = f_1\,d\lambda_1. \tag{5.18}$$

Similarly, for the equivalent simple shear, the only work done is that done by the shear stress t_{xy} (Fig. 5.3(a)), hence for the *equivalent* increase $d\gamma$ in shear strain

$$dW = t_{xy}\,d\gamma. \tag{5.19}$$

These two expressions must be identical, hence

$$t_{xy} = f_1 \frac{d\lambda_1}{d\gamma}.$$ (5.20)

Making use of the relation (5.9) between λ_1 and γ we obtain

$$\frac{d\lambda_1}{d\gamma} = \frac{1}{1 + \lambda_1^2}.$$ (5.21)

The shear stress in the equivalent simple shear thus becomes

$$t_{xy} = \frac{f_1}{1 + \lambda_1^2}.$$ (5.22)

Values of t_{xy} calculated in this way from the data presented in Fig. 5.8 are plotted in Fig. 5.9 against the equivalent shear strain γ

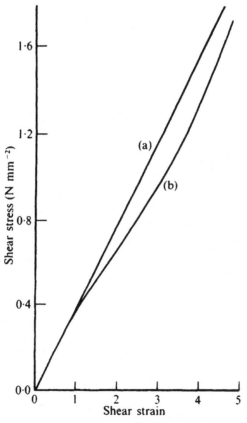

FIG. 5.9. Relation between shear stress and shear strain (curve (b)) calculated from data of Fig. 5.8. The theoretical line (a) has the slope $0 \cdot 39\,\text{N mm}^{-2}$.

derived from eqn (5.9). The linear relation (5.11) required by the theory, with a value of shear modulus G of $0·39\,N\,mm^{-2}$, is seen to fit the experimental data adequately up to a shear strain of $1·0$ (corresponding to an angle of shear ϕ of $45°$). Beyond this point the experimental curve falls below the theoretical line, in accordance with the corresponding deviation of the force in pure shear appearing in Fig. 5.8.

Alternative representation of experimental data

An alternative representation of the data of Figs 5.4, 5.5, and 5.8, which provides a more direct comparison between the various types of strain, is based on eqns (4.18), according to which the difference of principal stresses is proportional to the difference of the squares of the corresponding extension ratios. Conversion of the forces to

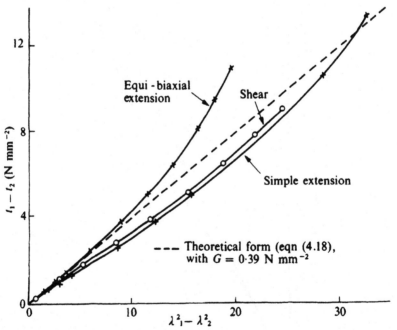

FIG. 5.10. Alternative representation of data given in figs. 5.3, 5.4 and 5.8.

true stresses yields the results shown in Fig. 5.10 for simple extension, equi-biaxial extension, and shear. It is seen that the three respective curves converge towards the theoretical straight line (of slope $0·39\,N\,mm^{-2}$) as the strain is reduced.

This form of plot brings out the considerable difference between the data for equi-biaxial extension (or equivalent uniaxial compression), which agree very closely with the theoretical form up to quite large values of the strain, and the data for extension and shear, both of which deviate markedly from the theory at relatively small values of strain. It also brings out the similarities in the final upward tendency in the region of very high strains, associated with the finite chain extensibility (non-Gaussian region).

5.4. Deviations from theory: Mooney equation

The deviations from the theoretical form of force–extension curve for simple extension have been extensively studied by later workers, who have shown that they are consistent with a semi-empirical formula devised by Mooney (1940), namely,

$$f = 2(\lambda - 1/\lambda^2)(C_1 + C_2/\lambda), \qquad (5.23)$$

where f, as before, is the force per unit unstrained area corresponding to the extension ratio λ. This formula contains two empirical constants, C_1 and C_2; the formula (5.3) derived from the statistical theory corresponds to the particular case $C_2 = 0$.

The theoretical basis of the Mooney equation is fully discussed in Chapter 10 and will not be considered at this stage except to say that it is derived on the assumption that Hooke's law is obeyed in simple shear. Eqn (5.23) is therefore the most general form of force–extension relation for simple extension consistent with a linear stress–strain relation in simple shear.

Writing eqn (5.23) in the form

$$\frac{f}{2(\lambda - 1/\lambda^2)} = C_1 + \frac{C_2}{\lambda}, \qquad (5.23a)$$

it is seen that a plot of $f/2(\lambda - 1/\lambda^2)$ against $1/\lambda$ should yield a straight line of slope C_2 and intercept $C_1 + C_2$ on the vertical axis at $1/\lambda = 1$. Typical plots of this kind for a series of vulcanized rubber compounds, taken from the work of Gumbrell, Mullins, and Rivlin (1953) are shown in Fig. 5.11, from which it appears that the constant C_1 varies widely according to the degree of vulcanization (the actual values ranging from $0 \cdot 10 \, \text{N mm}^{-2}$ to $0 \cdot 31 \, \text{N mm}^{-2}$), while C_2 remains approximately constant ($0 \cdot 10 \, \text{N mm}^{-2}$). This result suggests that C_1 is a function of the network structure and is

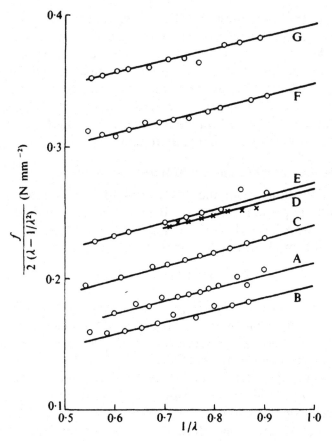

FIG. 5.11. Mooney plots for various rubbers in simple extension. (Gumbrell, Mullins, and Rivlin 1953.)

analogous to $\frac{1}{2}G$ in the statistical theory, while C_2 has some entirely independent origin. This tentative conclusion receives further support from the study of the effect of swelling by an organic liquid, which was originally shown by Gee (1946b) to reduce the extent of the deviations from the statistical theory. If the statistical theory were applicable we should have, on the basis of eqn (4.27) (p. 70),

$$f' = Gv_2^{-\frac{1}{3}}(\lambda - 1/\lambda^2),\qquad(5.24)$$

where f' is the force referred to the unstrained *unswollen* area, v_2 is the volume fraction of rubber, and $G = NkT$. The quantity ϕ, defined as

$$\phi = \frac{f' v_2^{\frac{1}{3}}}{2(\lambda - 1/\lambda^2)}, \tag{5.24a}$$

should on this basis be a constant. Fig. 5.12 shows plots of ϕ against $1/\lambda$, for different degrees of swelling, as obtained by Gumbrell,

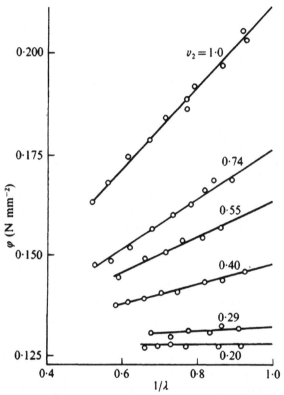

FIG. 5.12. Effect of swelling in benzene on behaviour in simple extension ($v_2 =$ volume fraction of rubber; ϕ defined by (eqn (5.24a)) (Gumbrell et al. 1953.)

Mullins, and Rivlin (1953). Their results are seen to be consistent with the modified Mooney formula

$$\frac{f' v_2^{\frac{1}{3}}}{2(\lambda - 1/\lambda^2)} = C_1 + \frac{C_2}{\lambda} \tag{5.25}$$

in which C_1 is approximately constant (independent of the degree of swelling) while C_2 falls progressively with increasing degree of swelling (decreasing v_2); at the highest degree of swelling ($v_2 = 0 \cdot 20$) the value of C_2 was found to be zero. The decrease in C_2 with increasing swelling indicates a systematic reduction and ultimate disappearance of the deviations from the statistical theory. It is remarkable also that these effects were found to be quite independent of the type of rubber (Fig. 5.13); other experiments showed

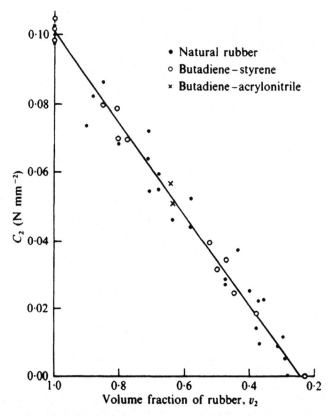

FIG. 5.13. Dependence of constant C_2 (eqn (5.25)) on v_2 for various rubbers. (Gumbrell, *et al.* 1953.)

that they were also independent of the nature of the swelling liquid. In all cases the value of C_2 tended to zero for a value of v_2 of about $0 \cdot 25$, corresponding to a volume swelling ratio of $4 \cdot 0$.

The general nature of these phenomena has been amply confirmed by numerous later workers, though with variations in detail.

As an example, reference may be made to the recent work of Allen, Kirkham, Padget, and Price (1971) on peroxide-cured natural rubber compounds. The only significant difference from the earlier work was in the values of C_2 for the unswollen rubbers; these were not entirely independent of the degree of cross-linking, but varied by a factor of about $2 \cdot 0$. A similar discrepancy was also obtained by Mullins (1959a) in a study of a wide range of peroxide- and sulphur-vulcanized rubbers.

The significance of the deviations from the statistical theory, as represented by the constant C_2 in the Mooney equation, cannot be assessed on the basis of experiments involving simple extension alone, but requires a consideration also of other types of strain. Studies of the stress–strain relations under the most general type of strain, i.e. the pure homogeneous strain, in which λ_1 and λ_2 may be independently varied, are discussed in Chapter 10. It is only on the basis of such studies that a general understanding of the quantitative significance of the deviations from the statistical theory can be obtained. In anticipation of the conclusions to be drawn from such studies, it is important to note here that the Mooney theory does *not* provide a satisfactory and self-consistent basis for the representation of the properties of a rubber in the most general state of strain, and that the values of the constants C_1 and C_2 (and in particular the ratio C_2/C_1) obtained from experiments on simple extension give a misleading and rather exaggerated impression of the magnitude of the deviations from the statistical theory.

The further question of the possible molecular interpretation of these deviations from the statistical theory must also be deferred until later, since this also requires the consideration of the data for all possible types of strain. Again in anticipation of the discussion of this subject in Chapter 10, it may be indicated here that no generally acceptable molecular interpretation of the observed effects has yet been established.

5.5. General conclusions

The conclusion to be drawn from the experimental observations reported in this chapter is that the formulae of the statistical theory, involving a single physical constant, correctly describe the properties of a real rubber to a first approximation. In particular, they provide a basis for the understanding of the relation between the stress–strain curves for widely different types of strain. However, in view of the very general nature of the theoretical arguments, and the simplifications necessarily introduced into the treatment of the network model, it is not surprising that some deviations from the ideal theoretical behaviour are to be found. These deviations, which other work has shown to be not limited to the particular case of natural rubber here examined, are of two kinds. First, at moderate strains, there is a tendency for the stresses to fall below the theoretical values, and secondly, at very large strains, the stresses tend to rise and may eventually greatly exceed the theoretical predictions. The first effect, which is not easy to understand, is discussed more fully in Chapter 10. The second is understood in principle; it arises from the finite extensibility of the network chains, and may be taken into account by a more complete statistical theory, as will be shown in the following chapter.

A further question of importance is the relation of the numerical value of the modulus (G in eqns (4.9a) and (4.9b)) to the number of chains per unit volume, which is determined by the degree of cross-linking. This question is examined in detail in Chapter 8.

6
NON-GAUSSIAN CHAIN STATISTICS AND NETWORK THEORY

6.1. Introduction

UP to this point the development of the statistical theory of the network has been based upon the approximate or 'Gaussian' distribution function (3.3) for the single chain, whose derivation assumes that the end-to-end distance r is very much less than the fully extended length of the chain R. The results derived on this basis are therefore limited to strains which are not too large, that is to say, to strains which do not begin to approach the limiting deformability of the network. In comparing the experimental stress–strain curves with the formulae derived from the Gaussian theory attention has already been drawn to the marked upturn in the stress in the region of very large strains, which is particularly evident in the case of simple extension (Fig. 5.4, p. 87). In this region, where an appreciable proportion of the chains become highly extended, the Gaussian statistical treatment is no longer valid, and a more accurate 'non-Gaussian' form of theory becomes essential.

The non-Gaussian statistical treatment of the single chain, which avoids the approximation inherent in the Gaussian theory, takes into account the finite extensibility of the chain, and thus leads to a more realistic form of distribution function which is valid over the whole range of r-values up to the maximum or fully extended length. In a similar manner, the treatment of the network of non-Gaussian chains leads to a limited extensibility of the network, the maximum extension being approximately proportional to the square root of the number n of random links in the chain. Hence as the chain length is reduced the maximum extension, and therefore also the range of validity of the Gaussian approximation, is correspondingly diminished. Since the chain length is inversely proportional to the degree of cross-linking, this implies that the Gaussian theory becomes increasingly inadequate as the degree of cross-linking is increased. For chains which are very short, e.g. corresponding to $n = 5$ or less, the mean chain extension even in the unstrained state already exceeds that for which the Gaussian approximation is valid; for networks of such short chains, therefore, a non-Gaussian treatment is essential for the accurate representation of the behaviour even under the smallest strains.

Unfortunately the attempt to replace the Gaussian statistical theory by a more exact treatment involves a considerable sacrifice of both simplicity and generality. As we have seen, the Gaussian statistical treatment is applicable to any type of molecular structure, and the Gaussian network has properties (such as, for example, the affine displacement of junction points) which enable it to be analysed, at least to a first approximation, in relatively simple mathematical terms. These advantages are no longer present with the non-Gaussian theory. Nevertheless, the subject is of such fundamental interest that a number of attempts to deal with it have been made and broad agreement has been reached on the general conclusions.

The following account begins with the non-Gaussian statistical treatment of the single chain—primarily the randomly jointed chain—and continues with the more difficult problem of the network of non-Gaussian chains.

6.2. Statistical treatment of randomly jointed chain

The inverse Langevin approximation

As in the elementary statistical theory dealt with in Chapter 3, the actual molecular structure is replaced by an idealized chain of n links, each of length l, such that the direction in space of any particular link, in the absence of external restraints, is entirely random and independent of that of neighbouring links in the chain. If we assume one end A of such a chain to be fixed at the origin of coordinates, as in Fig. 3.4 (p. 47), it is required to find the probability that the other end B shall be within a small volume element $d\tau$ in the neighbourhood of a point P at a distance r from the origin.

The method of solution of this problem given by Kuhn and Grün (1942) follows lines which are familiar in statistical thermodynamics. They first derive the most probable distribution of link angles with respect to the vector length AB. The probability of the given vector length is then taken to be simply the probability of this particular distribution of link angles.

For the calculation of the most probable distribution of link angles we note that the *a priori* probability that a particular link chosen at random shall be found in the angular interval θ_i to $\theta_i + d\theta_i$ is proportional to the corresponding solid angle, i.e. to

$$\tfrac{1}{2} \sin \theta_i \, d\theta_i.$$

The *a priori* probability of a distribution represented by n_1 links in the range $d\theta_1$, n_2 in the range $d\theta_2$, etc., *irrespective of order*, is therefore given by the expression

$$W_i = \prod_i (\tfrac{1}{2} \sin \theta_i)^{n_i} \frac{n!}{\prod(n_i!)}, \qquad (6.1)$$

in which the second factor represents the number of distinguishable permutations among the n_i. This probability has to be maximized for all possible variations in the n_i, subject to the conditions

$$\sum n_i = n, \qquad l \sum n_i \cos \theta_i = r. \qquad (6.2)$$

The solution, written in the form of a continuous function, is

$$dn = \frac{n\beta}{\sinh \beta} e^{\beta \cos \theta} \cdot \tfrac{1}{2} \sin \theta \, d\theta. \qquad (6.3)$$

This will be seen to involve only one parameter β, which is itself determined by the fractional extension of the chain r/nl, and is defined in the following way:

$$r/nl = \coth \beta - (1/\beta) = \mathcal{L}(\beta). \qquad (6.4)$$

The function $\mathcal{L}(\beta)$ is known as the Langevin function. Alternatively, we may write

$$\beta = \mathcal{L}^{-1}(r/nl), \qquad (6.5)$$

where $\mathcal{L}^{-1}$ is the corresponding *inverse* Langevin function.

The most probable distribution of link angles represented by (6.3) is thus determined only by the fractional extension r/nl, and is therefore independent of the actual value of n, the number of links in the chain. Its significance may be appreciated from the polar diagrams shown in Fig. 6.1, in which the radius vector is proportional to the number of links at the angle θ, omitting the purely geometrical factor $\tfrac{1}{2} \sin \theta$. A random distribution, corresponding to the unrestricted chain, would be represented on this type of plot by a circle. This is also the distribution for the case $r/nl = 0$. With increasing r/nl (fractional chain extension) the distribution becomes more and more asymmetrical.

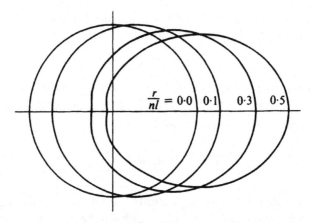

FIG. 6.1. Angular distribution of links in random chain. Parameter r/nl. (Kuhn and Grün 1942.)

The probability of a length r is assumed, as stated above, to be given by the probability of the most probable distribution of link angles. This is readily obtained by substituting the final distribution (6.3) into (6.1) and introducing Stirling's approximation

$$\ln n! = n \ln n - n \qquad (6.6)$$

for the factorials. In this way Kuhn and Grün obtain the *probability density* $p(r)$ in the logarithmic form

$$\ln p(r) = \text{constant} - n\left(\frac{r}{nl}\beta + \ln\frac{\beta}{\sinh\beta}\right). \qquad (6.7)$$

Comparison with Gaussian form

The form of (6.7) may be more readily appreciated from the series expansion, originally given by W. Kuhn and H. Kuhn (1946), i.e.

$$\ln p(r) = \text{constant} - n\left\{\frac{3}{2}\left(\frac{r}{nl}\right)^2 + \frac{9}{20}\left(\frac{r}{nl}\right)^4 + \frac{99}{350}\left(\frac{r}{nl}\right)^6 + \dots\right\}. \qquad (6.7a)$$

The corresponding Gaussian formula (eqn (3.5), p. 48) is equivalent to

$$\ln p(r) = \text{constant} - (3r^2/2nl^2), \qquad (6.7b)$$

which is seen to correspond to the first term in the series expansion

(6.7a). The Gaussian statistical theory is thus based on the assumption that all higher terms in r/nl are negligible compared with the first term.

Substitution of numerical values shows that for $r/nl = \frac{1}{3}$ the error in $\ln p(r)$ resulting from this assumption is about 3 per cent, while for $r/nl = \frac{1}{2}$ it amounts to about 8 per cent. For still larger chain extensions the Gaussian formula ceases to be a useful approximation.

The probability $P(r)$ of an end-to-end distance r irrespective of direction in space is obtained exactly as in the Gaussian case by multiplying the probability density $p(r)$ by the size of the volume element $4\pi r^2 \, dr$, i.e.

$$P(r) \, dr = 4\pi r^2 p(r) \, dr. \tag{6.8}$$

The form of the function $(1/4\pi r^2)P(r)$, which is equivalent to $p(r)$ in eqn (6.7), is shown on a logarithmic plot in Figs 6.2 and 6.3 for various values of n. The approach to the Gaussian form for small

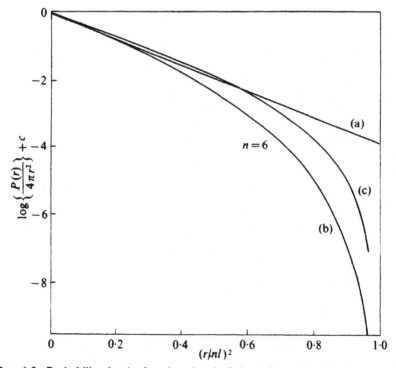

FIG. 6.2. Probability density functions for six-link random chain. (a) Gaussian; (b) inverse Langevin approximation (eqn (6.7)); (c) exact (eqn (6.13)).

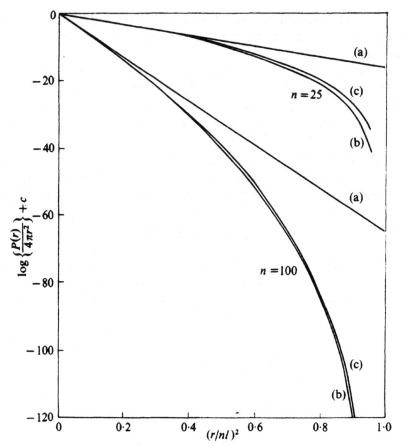

FIG. 6.3. Probability density functions for random chains of 25 and 100 links. (a) Gaussian; (b) inverse Langevin approximation (eqn (6.7)); (c) exact (eqn (6.13)).

values of r/nl is clearly shown, as is also the asymptotic approach to the maximum chain extension ($r/nl = 1$) at which the probability falls to zero ($\ln p(r) \to -\infty$).

6.3. Entropy and tension

The calculation of the entropy s for the single chain follows directly from the expression (6.7) for the probability density. Putting $s = k \ln p(r)$ we obtain the result

$$s = c - kn\left(\frac{r}{nl}\beta + \frac{\beta}{\sinh \beta}\right); \qquad \beta = \mathscr{L}^{-1}\left(\frac{r}{nl}\right),\dagger \qquad (6.9)$$

† This expression for the entropy has been obtained independently by J. J. Hermans (1943).

in which c is an arbitrary constant. The function within the bracket is identical with that in (6.7) and has the series expansion (6.7a).

The tension on the chain is derived exactly as in the Gaussian theory by differentiation of the entropy function. According to eqns (3.20) and (3.21) (p. 57) we have

$$f = -T\left(\frac{\partial s}{\partial r}\right).$$

Substitution of the expression (6.9) for s yields on differentiation

$$f = \left(\frac{kT}{l}\right)\mathscr{L}^{-1}\left(\frac{r}{nl}\right). \tag{6.10}$$

The function $\mathscr{L}^{-1}(r/nl)$ may also be expanded in series form to give

$$f = \frac{kT}{l}\left\{3\left(\frac{r}{nl}\right) + \frac{9}{5}\left(\frac{r}{nl}\right)^3 + \frac{297}{175}\left(\frac{r}{nl}\right)^5 + \frac{1539}{875}\left(\frac{r}{nl}\right)^7 + \ldots\right\}. \tag{6.10a}$$

As for the probability (and entropy), so also for the tension, the first term in the series corresponds to the Gaussian approximation (cf.

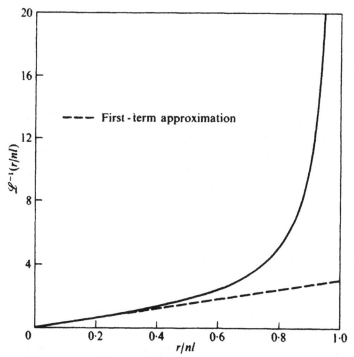

FIG. 6.4. Form of force–extension relation for a random chain (eqn (6.10)).

eqns (3.19) and (3.22)). The form of the complete function (6.10) is shown in Fig. 6.4, and is seen to approximate to linearity for values of fractional chain extension less than about $\frac{1}{3}$.

6.4. Alternative derivation of tension on chain

An alternative treatment of the problem, given originally by James and Guth (1943) and later by Flory (1953), enables the tension on the chain to be derived directly, without proceeding by way of the entropy.

According to this method we consider the chain to be fixed at one end (Fig. 6.5), and to be acted on at the other end by a force f in a

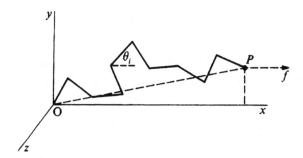

FIG. 6.5. Derivation of the tension on a single chain.

fixed direction, which may be taken to be parallel to the x-axis. The total chain length will then be the sum of the x-components for the individual links; hence to obtain the mean chain length it is sufficient to find the mean value of the x-component for all the links. Since a link inclined at an angle θ_i to the x-axis has a component of length $x_i = l \cos \theta_i$, the rotation of a particular link from the position corresponding to $\theta = 0$ to the position $\theta = \theta_i$ (with all other link angles held constant) therefore results in the performance of work against the applied force amounting to $fl(1 - \cos \theta_i)$. The contribution of this particular link to the orientational potential energy may thus be taken as $-fl \cos \theta_i$ or $-fx_i$ (neglecting the additive constant fl, which is the same for all links). Introducing the Boltzmann concept, the probability of the particular link angle may then be assumed to be proportional to $\exp (fx_i/kT)$; the mean value of x_i

will thus be given by

$$\bar{x}_i = \frac{\int_{-l}^{l} x_i \exp\left(fx_i/kT\right) dx_i}{\int_{-l}^{l} \exp\left(fx_i/kT\right) dx_i} = L\mathscr{L}\left(\frac{fl}{kT}\right),$$ (6.11)

in which $\mathscr{L}$ is the Langevin function defined by (6.4). This is the mean component of length for an arbitrarily chosen link. Since the same formula applies to all the links in the chain, the mean value of x for the whole chain is simply $n\bar{x}_i$. Since, by symmetry, the mean y- and z-components are zero, the mean r vector coincides with the x-axis. Hence we may write

$$r = n\bar{x}_i = nl\mathscr{L}\left(\frac{fl}{kT}\right),$$ (6.12)

$$f = \left(\frac{kT}{l}\right)\mathscr{L}^{-1}\left(\frac{r}{nl}\right),$$ (6.12a)

which is identical with (6.10) above.

6.5. The exact distribution function

The preceding treatments both utilize methods commonly employed in statistical thermodynamics. It is important to remember, however, that these methods yield exact results only when the number of units in the statistical assembly is sufficiently large. In most molecular systems, where we are concerned with numbers of the order of 10^{23}, this condition is amply satisfied. In the present problem, however, the number of links in the chain is by no means large in this sense and the accuracy of the results obtained by these methods must become increasingly open to question as this number is reduced. It is not easy to decide a priori precisely how much error is to be expected for any particular value of n. Fortunately, however, an alternative method of deriving the distribution function in exact mathematical terms, which does not involve statistical-thermodynamic arguments, is available. This method applies for any value of n (from 1 upwards) and may therefore be used to assess the accuracy of the more general methods in the case when n is small.

The alternative method treats the problem as a purely geometrical question, and involves no approximations of any kind. The solution was originally obtained by the author (Treloar 1946) by a simple transformation of a result previously obtained by Hall (1927) and by Irwin (1927) in connection with the theory of random sampling, and has since been derived independently by Wang and Guth (1952) and Flory (1969, p. 313). For a chain of n links, each of length l, the probability of an end-to-end distance between r and $r+dr$ (irrespective of direction in space) is represented by the formula

$$P(r)\,dr = \frac{r}{2l^2}\frac{n^{n-2}}{(n-2)!}\sum_{s=0}^{k}(-1)^s\times{}^nC_s\{m-(s/n)\}^{n-2}\,dr \quad (6.13)$$

where $k/n \leq m \leq (k+1)/n$, $m = \frac{1}{2}(1-r/nl)$, and ${}^nC_s$ represents the number of combinations of n things taken s at a time.

Distribution functions for three different values of n, calculated from this formula, are shown in Figs 6.2 and 6.3. The quantity plotted is, as before, the probability density $(1/4\pi r^2)P(r)$, an arbitrary constant having been added so that all the curves start from the origin. It will be seen that the curves for different n-values are not all of identical form, i.e. the probability is not simply a function of r/nl, as it is for the inverse Langevin approximation. Comparison of the respective curves shows that for $n=6$, the latter, though better than the Gaussian form, is quantitatively seriously in error. At $n=25$, the difference between the two formulae is already quite small, while at $n=100$ it is practically negligible. The logarithmic plot, being directly related to the entropy of the chain (eqn (6.9)), provides the most direct representation of the significance of the non-Gaussian statistical theory in relation to the problem of rubber elasticity. For other applications, however, a linear plot may be more appropriate. On a linear plot (Fig. 6.6) the differences between the various formulae are much less noticeable, even for values of n as low as 6.

In an actual rubber the effective value of n will depend not only on the number of bonds in the chain, but also on the chain 'stiffness', which will be a function of its chemical constitution. For natural rubber values of n (for the chains in a normal cross-linked network) are likely to lie in the range 50–100 (cf. § 6.10 below) and the use of the inverse Langevin approximation will therefore be justified. For certain other systems (e.g. cross-linked polythenes), however, very

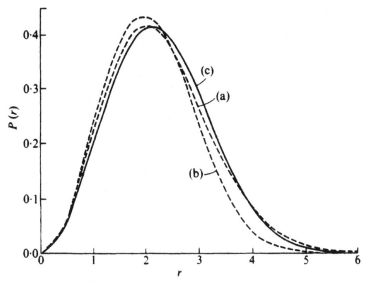

FIG. 6.6. Distribution functions $P(r)$ for six-link random chain, on linear plot; $l = 1$.
(a) Gaussian; (b) inverse Langevin; (c) exact.

much smaller effective chain lengths—corresponding to $n = 10$ or less—may occur, and this approximation may not be sufficiently exact. Whether, apart from its mathematical complexity, the exact distribution is in such cases to be preferred is questionable, since an actual 'stiff' chain cannot be adequately replaced by a randomly jointed chain containing a small number of links. The distinction between the different treatments of the random chain is therefore somewhat academic, and is probably less important than the question of the relation of the randomly jointed chain model to any actual molecular structure.

6.6. Application to real molecular structures

This question of the relation of the randomly jointed chain to an actual molecular structure raises problems of very considerable difficulty. In an actual molecule successive bonds are not unrelated in direction but are joined together at a specific angle (valence angle). In addition to this, there will be interactions of an energetic character between neighbouring atoms in the chain which impede the rotation to a greater or less extent. While it may be shown quite generally (Flory 1969, p. 6) that the Gaussian distribution represents the limiting form of distribution for *any* chain structure of sufficient length, provided that the condition $r \ll R$ (where R is the fully extended length) is satisfied, no comparable generality can be shown to exist in the non-Gaussian region, where this condition does not apply. Each particular chain structure must therefore be examined independently, using numerical methods of computation.

In rather crude studies of this kind, the author showed that for *freely rotating* models of both the polymethylene (polythene) and rubber structures the distribution function for long chains approximated to the inverse Langevin form (6.7) derived for the randomly jointed chain, with suitably chosen values of n, the number of random links (Treloar 1943c, 1944b). This is shown in Fig. 6.7. In

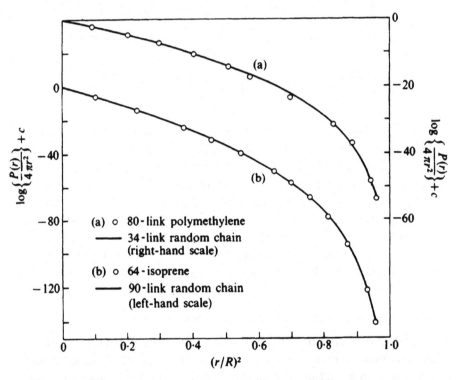

FIG. 6.7. Comparison of freely rotating models of polymethylene and rubber chains with randomly jointed chain.

the case of the polymethylene chain the number of C—C bonds per random link, namely, 80/34 (or 2·35), differs somewhat from the figure of 3·0 bonds per random link obtained for the same model in the Gaussian region (eqns (3.15c), p. 54). Similarly for the poly-isoprene chain the figure of 64/90 (or 0·71) isoprene units per random link is not quite the same as the value 0·77 isoprene units for the Gaussian region (p. 55). These differences, however, are not thought to be significant, in view of the rather large errors inherent in the method used in the calculations, and it may therefore be concluded that the distribution functions for typical chain structures approximate to the form derived for the randomly jointed chain, and are relatively insensitive to the details of the structure.

This degree of generality encourages us to expect that the same conclusion might be extended to more realistic types of chain in which hindrances to rotation and other complications may be present. Some direct evidence to this effect is provided by the recent work of Hill and Stepto (1971) in which interactions between neighbouring groups in the polymethylene-type chain are included. Their calculations indicate an increasingly close approach to the random-chain distribution, as represented by the exact formula (6.13), as the number of bonds is increased from small values up to 30.

6.7. Non-Gaussian network theory

As was noted at the beginning of this chapter the various simplifications which so greatly facilitate the mathematical treatment of the properties of the Gaussian network are no longer justifiable in the vastly more complex case of the network of non-Gaussian chains. The treatment of this problem in a rigorous manner presents formidable mathematical difficulties, which can be reduced to manageable proportions only by the introduction of assumptions which are not strictly valid. To this extent the conclusions arrived at must be regarded as somewhat uncertain in the quantitative sense.

Simplified theory: three-chain model

The simplest and by far the most tractable model of the non-Gaussian network, which nevertheless brings out all its essential characteristics, is the three-chain model. This is based on the assumption, strictly valid only for Gaussian chains, that the network may be replaced by three independent sets of chains parallel to the axes of a rectangular coordinate system. It is to be expected that this assumption, though not strictly justified, will not be too seriously in error, except possibly at extreme deformations.

Let us assume that in the unstrained state the r vectors of the three chains lie along the $0x$, $0y$, and $0z$ axes, respectively, and that each of these has the value r_0. It is further assumed that the effect of the deformation is to change the r vectors in the same ratio as the corresponding dimensions of the bulk rubber, i.e. that the chains undergo an *affine* deformation. For a simple extension in the ratio λ in the direction $0x$ the r vectors of the three chains in the deformed state, assuming constancy of volume, will be

$$r_x = \lambda r_0, \qquad r_y = r_z = r_0/\lambda^{\frac{1}{2}}. \tag{6.14}$$

We now introduce the non-Gaussian entropy function (6.9) to represent the entropies s_x, s_y, and s_z of the three chains. Since $s_y = s_z$ the total entropy in the deformed state will accordingly be given by

$$s_x + 2s_y = -kn\left\{\frac{r_0\lambda}{nl}\mathscr{L}^{-1}\left(\frac{r_0\lambda}{nl}\right) + \ln\frac{\mathscr{L}^{-1}(r_0\lambda/nl)}{\sinh \mathscr{L}^{-1}(r_0\lambda/nl)}\right\}$$
$$-2kn\left\{\frac{r_0\lambda^{-\frac{1}{2}}}{nl}\mathscr{L}^{-1}\left(\frac{r_0\lambda^{-\frac{1}{2}}}{nl}\right) + \ln\frac{\mathscr{L}^{-1}(r_0\lambda^{-\frac{1}{2}}/nl)}{\sinh \mathscr{L}^{-1}(r_0\lambda^{-\frac{1}{2}}/nl)}\right\}. \tag{6.15}$$

If N is the number of chains per unit volume, the total entropy S per unit volume will be $(N/3)(s_x + 2s_y)$. If S_0 is the entropy in the unstrained state the work of deformation then becomes

$$W = -T(S - S_0) = -T\{(N/3)(s_x + 2s_y) - S_0\}. \tag{6.16}$$

The tensile force f per unit unstrained area is then obtained by differentiation, i.e.

$$f = \frac{dW}{d\lambda} = -\frac{NT}{3}\frac{d(s_x + 2s_y)}{d\lambda}. \tag{6.17}$$

Substitution of the expression (6.15) in (6.17) followed by differentiation with respect λ yields

$$f = \frac{NkT}{3}\frac{r_0}{l}\left\{\mathscr{L}^{-1}\left(\frac{r_0\lambda}{nl}\right) - \lambda^{-\frac{3}{2}}\mathscr{L}^{-1}\left(\frac{r_0\lambda^{-\frac{1}{2}}}{nl}\right)\right\}. \tag{6.18}$$

This may be further simplified if r_0, the vector chain length in the unstrained state, is given the r.m.s. value for the free chain, namely, $ln^{\frac{1}{2}}$. The final result is

$$f = \frac{NkT}{3}n^{\frac{1}{2}}\left\{\mathscr{L}^{-1}\left(\frac{\lambda}{n^{\frac{1}{2}}}\right) - \lambda^{-\frac{3}{2}}\mathscr{L}^{-1}\left(\frac{1}{\lambda^{\frac{1}{2}}n^{\frac{1}{2}}}\right)\right\}. \tag{6.19}$$

This relation between the force and the extension ratio, which was originally obtained by James and Guth (1943) by an essentially similar method, reduces to the Gaussian formula for values of λ which are not too large. For on substitution of the first term in the expansion for $\mathscr{L}^{-1}$ given in eqn (6.10a), i.e. $\mathscr{L}^{-1}(x) \simeq 3x$, into eqn (6.19) the result is

$$f = NkT(\lambda - 1/\lambda^2), \tag{6.19a}$$

in agreement with the expression derived in Chapter 4 (eqn (4.19a)).

The corresponding relations for the difference of principal stresses in a pure homogeneous strain of the most general type, defined by principal extension ratios λ_1, λ_2, and λ_3 are easily shown to be of the form (cf. Wang and Guth 1952)

$$t_1 - t_2 = \frac{NkT}{3} n^{\frac{1}{2}} \left\{ \lambda_1 \mathscr{L}^{-1}\left(\frac{\lambda_1}{n^{\frac{1}{2}}}\right) - \lambda_2 \mathscr{L}^{-1}\left(\frac{\lambda_2}{n^{\frac{1}{2}}}\right) \right\}, \tag{6.20}$$

in which t_1 and t_2 are the principal stresses. Again, on insertion of the small-strain approximation $\mathscr{L}^{-1}(x) \simeq 3x$, this reduces to the previously obtained Gaussian relation (eqns (4.18)),

$$t_1 - t_2 = NkT(\lambda_1^2 - \lambda_2^2). \tag{6.20a}$$

The form of the force–extension curves corresponding to eqn (6.19) is shown in Fig. 6.8 for three values of n. The strong upward curvature at high extensions is the direct consequence of the limited extensibility of the chains, as can be seen by comparing these curves with the force–extension curve for the single chain (Fig. 6.4). The maximum extension ratio for the network, like that for the single chain, is equal to the square root of the number of random links in the chain. The non-Gaussian force–extension relations are thus determined by two parameters, of which the first, N, defines the vertical scale or modulus in the region of small or moderate strains, while the second, n, which is specific to the non-Gaussian theory, controls the behaviour in the high-strain region and the ultimate extensibility of the network. Mathematically, these two parameters may be treated as independent variables, though in physical reality they are not independent of each other. For as we have seen, N is related to the chain molecular weight M_c (eqn (4.9c), p. 65) and for any given polymer the value of M_c automatically determines the number n of equivalent random links; as a result n must be

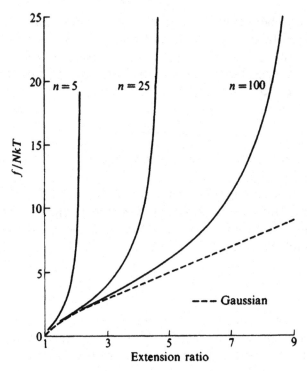

FIG. 6.8. Non-Gaussian force–extension curves from three-chain model (eqn (6.19)).

inversely proportional to N. In general, however, the value to be assigned to the equivalent n for any given M_c value is not known, so that the numerical relationship between N and n cannot be given in general terms.

The four-chain (tetrahedral) model

A much more realistic representation of the conditions existing in an actual network may be obtained by a suitable modification of the 4-chain model originally introduced by Flory and Rehner (1943) for the development of the Gaussian theory. This model has the further advantage of being adaptable also to the problem of calculating the photoelastic properties of the non-Gaussian network, discussed in Chapter 9.

The model considers an elementary 'cell' of the network consisting of four chains radiating outwards from a common junction point P (Fig. 6.9). If the chains all have the same contour length, the

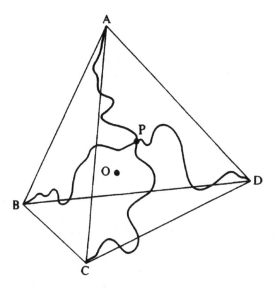

FIG. 6.9. The Flory–Rehner 'tetrahedral' model.

average positions of their outer ends (i.e. of the nearest neighbour junction points) will be at the four corners of a regular tetrahedron ABCD. In Flory and Rehner's treatment of the Gaussian network the outermost junction points were considered to be fixed at their mean positions, while the central junction point was allowed to fluctuate in position in response to the random thermal motion of the associated chains. The deformation is assumed to displace the relative positions of the outermost junction points in a manner corresponding to the macroscopic strain (affine deformation) but the central junction point is allowed to fluctuate, as in the unstrained state. The entropy of formation of the four-chain cell of the network from the original independent chains was calculated first for the unstrained state and then for the strained state, the difference of entropy for these two states being taken as the entropy of network deformation.

An essential part of the calculation of the entropy of formation involves finding the probability that the four chains, whose outer junction points are fixed at A, B, C, and D, shall meet within a small volume element in the neighbourhood of some arbitrarily chosen point P. Integration of this probability over all possible positions of P then gives the total probability of their meeting at any point in space.

So long as the calculation is restricted to the Gaussian region of extension, certain terms in the calculation do not affect the resultant entropy of deformation. In particular it is found, first, that the most probable position of the central junction point is at the centre of the deformed tetrahedron, i.e. at the position corresponding to an affine displacement of the centre O of the original cell. Secondly, the fluctuations of position of the central junction point (which are themselves described by a spherically symmetrical Gaussian distribution) are found to be independent of the state of strain and therefore do not enter into the calculation of the entropy of deformation. It follows from this that (in the Gaussian case) it makes no difference to the result whether the central junction point is treated as fluctuating or as fixed at its most probable position, corresponding to an affine deformation.

These conclusions illustrate some of the general properties of the Gaussian network already discussed in Chapter 4. None of these conclusions apply to the non-Gaussian network. A realistic consideration of the corresponding non-Gaussian model should therefore take into account a number of additional factors which do not arise in the case of the Gaussian theory. Of these, the three most important are the following:

(1)   the non-Gaussian character of the individual chains;
(2)   the non-affine displacement of the mean or equilibrium position of the central junction point;
(3)   the variation of the extent of the fluctuations of the central junction point with strain.

Unfortunately the detailed calculations of the entropy of deformation, taking account of all these factors, can be carried out only by numerical methods of computation. Such calculations have been carried out by the author (Treloar 1946) on the basis of the exact series expression for the distribution of r-values for the randomly jointed chain discussed in § 6.5. These calculations showed that for n-values of 25 and 100 the contribution to the entropy arising from the fluctuations of the central junction point (item (3) above) was negligible compared with that due to the deformation of the chains.† In later work (Treloar 1954), based on the inverse Langevin approximation, this factor was therefore neglected, and attention was concentrated on the effect of the non-affine displacement

† This is consistent with the more general conclusion of Wang and Guth (1952) that the neglect of fluctuations of junction points leads to an error of the order $1/n$.

of the central junction point. A typical result (Fig. 6.10), in which calculations based on (a) affine and (b) non-affine displacements are compared, shows the latter to have a significant effect on the extensibility of the network at high values of the stress.

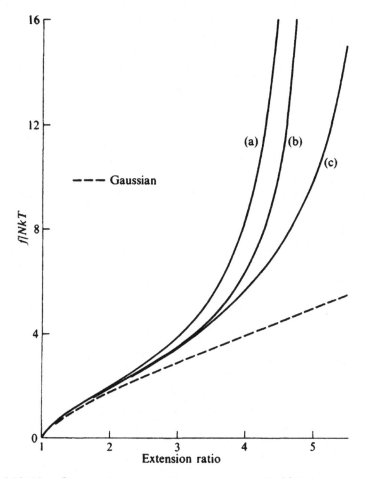

FIG. 6.10. Non-Gaussian force–extension curves for $n = 25$. (a) 3-chain model; (b) tetrahedral model, affine displacement; (c) tetrahedral model, non-affine displacement.

Inverse Langevin series approximation

One disadvantage of the preceding model is that in the non-Gaussian region the elementary four-chain cell ceases to behave isotropically, i.e. the properties depend to some extent on the *direction* of the applied strain. In the calculations referred to above the results quoted represented the *mean* properties for three mutually perpendicular directions of straining. The same difficulty arises in a more acute form with the three-chain model of James and Guth, but in this case the strain was arbitrarily considered to be applied along one of the chain directions. In any such simplified 'cell' model the properties necessarily reflect the particular geometry of the chosen system. To avoid this difficulty it is desirable to operate on the total assembly of chains, which in the unstrained state corresponds to a random angular distribution of *r* vectors in space. Moreover, by using the inverse Langevin expression for the chain entropy in the series form (eqn (6.7a)) it is possible to derive an explicit algebraic solution to the problem in the form of a corresponding series, which in principle can be developed to any required degree of accuracy by the inclusion of a suitable number of terms. This approach has been discussed by Wang and Guth (1952), and developed more fully by the author (Treloar 1954). According to the latter, if the r.m.s. length in the unstrained state is r_0, the number of chains having *r* vectors in the range *r* to $r + dr$ after a deformation defined by the extension ratio λ, assuming an affine deformation of junction points, is given by

$$f(r)\,dr = \left\{ \frac{N}{r_0(\lambda - 1/\lambda^2)^{\frac{1}{2}}} \cdot \frac{r}{(r^2 - r_0^2/\lambda)^{\frac{1}{2}}} \right\} dr. \qquad (6.21)$$

Introduction of the value of the chain entropy given by the series expansion (6.7a) for this group of chains, followed by integration over all chains, yields the total entropy of the system, from which the force is obtained by differentiation with respect to λ. The result is expressed in the form of a series, of which the first five terms are

$$f = NkT\left(\lambda - \frac{1}{\lambda^2}\right)\left\{1 + \frac{3}{25n}\left(3\lambda^2 + \frac{4}{\lambda}\right) + \frac{297}{6125n^2}\left(5\lambda^4 + 8\lambda + \frac{8}{\lambda^2}\right) + \right.$$

$$+ \frac{12\,312}{2\,205\,000n^3}\left(35\lambda^6 + 60\lambda^3 + 72 + \frac{64}{\lambda^3}\right) +$$

$$+ \frac{126\,117}{693(673\,750)n^4}\left(630\lambda^8 + 1120\lambda^5 + 1440\lambda^2 + \frac{1536}{\lambda} + \frac{1280}{\lambda^4}\right) +$$

$$\left. + \dots \right\}. \qquad (6.22)$$

This expression yields the result represented in Fig. 6.11 for the case $n = 25$ (curve (a)).

The above method becomes difficult to apply at strains approaching the limiting extension of the network, owing to the excessive number of terms, of rapidly increasing complexity, which are required. To check the accuracy of the five-term approximation, the author compared the above curve with the result obtained by a modified treatment in which the complete function (6.7) for the chain entropy was used, the summation over all chains being carried out by graphical integration. This result, shown by curve (b) in Fig. 6.11, is in close agreement with that obtained from the four-chain model, assuming an affine deformation of the central junction point

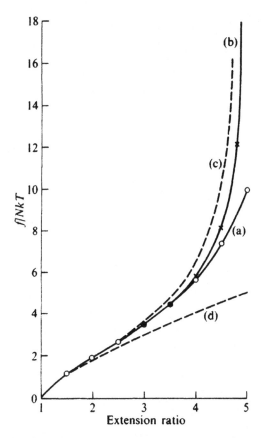

FIG. 6.11. Non-Gaussian force–extension curves, $n = 25$. (a) Complete assembly of chains, series expansion (6.22); (b) complete assembly of chains, graphical integration; (c) four-chain model (as curve (b) in Fig. 6.10); (d) Gaussian.

(curve (c)), but differs substantially from the five-term series expansion at extensions near to the maximum. It is therefore concluded that the inclusion of five terms in the series is inadequate for a proper representation of the behaviour in this region.

6.8. Comparison with experiment

In view of the rather slight differences between the results derived from the various forms of non-Gaussian network theory, the choice of a model for comparison with experimental data may well be determined by ease of computation rather than theoretical precision. On this score, the three-chain model of James and Guth possesses overwhelming advantages.

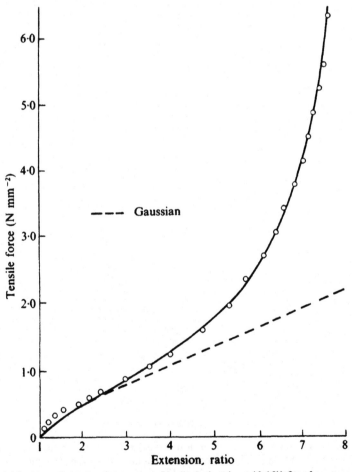

FIG. 6.12. Non-Gaussian force–extension relation (eqn (6.19)) fitted to experimental data, with $NkT = 0.273$ N mm^{-2}, $n = 75$.

Fig. 6.12 represents an attempt to fit the experimental data for simple extension (already given in Fig. 5.4, p. 87) by the non-Gaussian formula (6.19), by suitable choice of the parameters N and n. Owing to the deviations from the (Gaussian) statistical theory in the region of low extensions, discussed in the previous chapter, it is not possible to obtain a satisfactory fit to the data over the whole range of extension. The curve shown in Fig. 6.12 has been adjusted to give a good fit in the high-extension region ($\lambda > 3$), but it is rather inadequate in the region of low strains. There is no doubt, however, that the non-Gaussian theory gives a very much better representation of the general features of the experimental curve than the Gaussian theory, which entirely fails to represent the point of inflection and subsequent rapidly increasing slope at high strains.

6.9. Possible influence of crystallization

In considering the application of the non-Gaussian theory to experimental data, no account has been taken of the possible effects of strain-induced crystallization on the mechanical properties of the rubber. It so happens that the region of strain in which the non-Gaussian theory becomes relevant coincides roughly with the region where crystallization develops most strongly (cf. Fig. 1.9, p. 21), and it has been suggested (Flory 1947) that the upturn in the force–extension curve should be attributed primarily to this factor. Certainly the crystallization which occurs would be expected to produce some stiffening of the rubber, but it is not easy to predict the magnitude of the effect. It has been noted, however, by Wang and Guth (1952) that the characteristic form of the force–extension curve for natural rubber is only slightly affected by raising the temperature to 100 °C, though this will substantially reduce the crystallinity (cf. Fig. 9.7, p. 190). They note also that the characteristic inflection is also present in the case of non-crystallizing synthetic rubbers.

The later work of Smith, Greene, and Ciferri (1964), using X-ray diffraction and other methods of detecting the onset of crystallinity, has shown quite definitely that the initial upturn in the force–extension curve is a genuine non-Gaussian effect, unrelated to crystallization, though at higher extensions complications associated with crystallization were observed.

In view of all the evidence, therefore, it seems reasonable to interpret the main features of the experimental force–extension curve in terms of the non-Gaussian statistical theory, and to regard the effects of crystallization as secondary in character, producing only minor modifications.

6.10. The equivalent random link

From the values of the parameters N and n required to fit the experimental force–extension curve in Fig. 6.12 it is possible to estimate the number of isoprene units in the equivalent random link. The value $NkT = 0.273 \, \text{N mm}^{-2}$ corresponds to a 'chain' molecular weight M_c (eqn (4.9c), p. 65) of $8.31 \, \text{kg mol}^{-1}$; since this is equivalent to 75 random links, it follows that the 'molecular weight' of the equivalent random link is $0.111 \, \text{kg mol}^{-1}$, which is equal to 1.63 isoprene units.

As would be expected, this figure is considerably in excess of the value 0.77 isoprene units per random link derived theoretically on the basis of a freely rotating chain model, and indicates a considerable 'stiffness' due to energetic interactions within the chain.

The above result, however, is based on a rather crude method of fitting the experimental curve, and cannot be regarded as highly accurate. In an attempt to overcome this difficulty a semi-empirical approach has been proposed in which a 'Mooney' term is incorporated into the non-Gaussian theory. The first such attempt was that of Mullins (1959b) who presented his data for a series of vulcanizates in the form of 'Mooney' plots, as in Fig. 6.13. These curves are essentially similar to those of Gumbrell et al. (1953) given in Fig. 5.11 (p. 96), but are continued into the non-Gaussian region of strain. Assuming that the linear portion of these curves corresponds with the Gaussian region, as modified by an additional Mooney C_2 term, the deviations from linearity with increasing strain (decreasing $1/\lambda$) may be taken to represent the non-Gaussian terms in the series expansion (6.22). To obtain the appropriate value of n (the number of equivalent random links in the chain), Mullins first calculates C_1 and C_2 from the linear region, then finds the value of λ, i.e. $\lambda = \lambda^*$, at which the curves deviate from the straight line by an arbitrary amount (2.5 per cent of the C_1 term). Since only a small contribution from the non-Gaussian terms is therefore involved, it is sufficient, in relating λ^* to n, to retain only the first three terms in the expansion (6.22). For a given value of n, λ^* is then the value of λ for which the function

$$1 + \frac{3}{25n}\left(3\lambda^2 + \frac{4}{\lambda}\right) + \frac{297}{6125n^2}\left(5\lambda^4 + 8\lambda + \frac{8}{\lambda^2}\right) \qquad (6.22a)$$

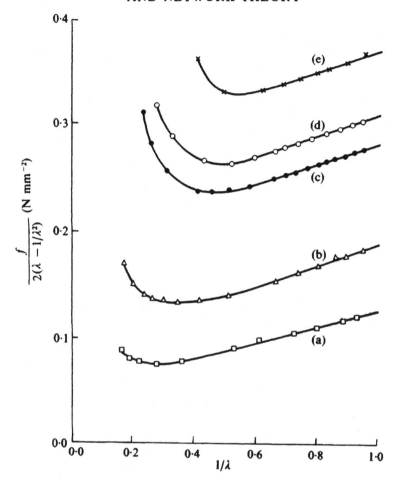

FIG. 6.13. 'Mooney' plot of experimental force–extension data for peroxide cross-linked rubbers containing (a) 1 per cent; (b) 2 per cent; (c) 3 per cent; (d) 4 per cent; (e) 5 per cent peroxide. (Mullins 1959b.)

is equal to 1·025 times its value at $\lambda = 1$. In this way Mullins originally obtained the figure of 1·1 isoprene units for the equivalent random link. Unfortunately, his calculations contained a numerical error; the corrected relation between λ^* and n is given in Table 6.1.

TABLE 6.1

Conjugate values of λ^* and n which give 2·5 per cent increase in C_1 term (Mullins, private communication)

n	50	100	150	200	250	300
λ^*	2·21	2·90	3·43	3·89	4·30	4·68

Introduction of this correction increases the figure for the number of isoprene units per random link from 1·1 to 1·5. This revised figure is not very different from the value 1·63 obtained from the more elementary method of curve-fitting.

The later work of Morris (1964), incorporating a more complete analysis, has raised doubts about the reliability of the Mullins method. Starting from an identical theory (i.e. using the first three terms in eqn (6.22) together with a Mooney C_2 term), Morris shows that the value of n deduced from λ^* provides a very inadequate basis for predicting the form of the curve at high extensions, where the non-Gaussian terms become predominant. He therefore finds the best fit over the whole range of strain, treating C_1 (i.e. NkT), C_2, and n as adjustable parameters. Fig. 6.14 shows that the fit so

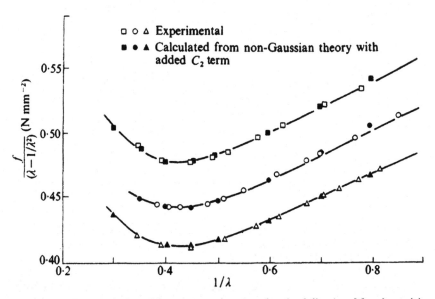

FIG. 6.14. 'Mooney' plot of force–extension data for the following M_c values: (a) 5·478 kg mol^{-1}; (b) 5·505 kg mol^{-1}; (c) 5·644 kg mol^{-1}. (Morris 1964.)

obtained is very good. His results, however, yield a very much higher value for the equivalent random link, namely 4·3 isoprene units.

There is no doubt that, on the basis of the chosen model, the treatment of Morris is the more accurate. His figure for the equivalent random link, however, seems unduly high, when compared either with the admittedly crude figure obtained from the curve in Fig. 6.12 or with the values obtained from photoelastic data (Chapter 9). The Mooney equation itself, however, is open to serious criticism as a means of representing the general stress–strain behaviour of rubber (Chapter 10), and the further identification of the constant C_1 with NkT in the statistical theory cannot be regarded as soundly based. In the circumstances, therefore, the significance of any conclusions drawn from this type of analysis must be regarded as open to question.

7

SWELLING PHENOMENA

7.1. Introduction

A NUMBER of references have already been made to the swelling of rubbers by organic liquids and to the effects of such swelling on their mechanical properties. The nature of the swelling process in itself, and the factors which determine the degree of swelling attained in any particular circumstances, have not, however, yet been considered.

The property of swelling in suitable low-molecular-weight liquids is one which is possessed by a wide range of high polymers, both natural and synthetic. In many respects this property is akin to solution, and, as in the case of solution, it is markedly dependent on the nature of the swelling liquid. Just as materials may be devided in respect of solubility into those which are soluble in water (hydrophilic) and those which are soluble in organic—e.g. hydrocarbon—solvents (hydrophobic), so polymers may similarly be divided in respect of swelling into the water-swelling and the organic-liquid-swelling classes. The first class includes cellulose (cotton, wood, etc.), proteins (gelatin, wool, silk, etc.), while the dominant group in the second class consists of the rubbers, both natural and synthetic. Fortunately for the present purpose the rubbers as a class are somewhat simpler in their swelling behaviour than the water-absorbing systems in the sense that their behaviour may be described at least to a first approximation in terms of a rather simple statistical-thermodynamic model; a correspondingly general model does not appear to be possible in the case of water-absorbing materials.

For a complete exposition of swelling phenomena it would be desirable to incorporate a general account of solution phenomena in low-molecular-weight materials as well as in high-polymer systems. This, however, is a large subject, and would be beyond the scope of the present work. The treatment here must be limited to the rather small section of this subject which has a direct bearing on the phenomena of swelling in rubber-like materials; the reader who wishes to pursue the subject further is referred to the excellent texts by Hildebrand and Scott (1950), Guggenheim (1952), and Tompa (1956).

7.2. General thermodynamic principles

In swelling, as in solution, we are concerned with the equilibrium between phases. The simplest case to consider is that in which one phase contains two components, while the other contains only one. In normal solutions the two-component or mixed phase is liquid while the one-component or pure phase is solid; in the case of swelling the situation is reversed, the mixed phase being the solid and the pure phase the liquid. From the thermodynamic standpoint, however, this difference has no significance.

The equilibrium of any system of this type is determined by the condition that its free energy shall be a minimum with respect to changes in the composition of the mixed phase. Thus for the particular case when the mixed phase is polymer plus liquid and the single phase is the pure liquid, this means that the change in free energy resulting from the transfer of a small quantity of liquid from the pure liquid phase to the mixed phase shall be zero. To represent this quantitatively it is convenient to introduce the *Gibbs free energy of dilution* ΔG_1, defined as the change in the Gibbs free energy of the system due to the transfer of unit quantity (1 mol) of liquid (component 1) from the liquid phase to a very large quantity of the mixed phase. For a system at constant pressure the condition for equilibrium with respect to the transfer of liquid is then

$$\Delta G_1 = 0. \tag{7.1}$$

The total free energy of dilution may be expressed in terms of the heat of dilution ΔH_1 and the entropy of dilution ΔS_1. Thus

$$\Delta G_1 = \Delta H_1 - T\,\Delta S_1. \tag{7.2}$$

In this expression ΔH_1 and ΔS_1 are the corresponding changes in the heat content H and entropy S of the system per mole of liquid transferred from the liquid phase to the mixed phase. The heat content H is defined by the relation

$$H = U + pV, \tag{7.3}$$

in which U is the internal energy and V the volume. When p is the atmospheric pressure, the effect of the term pV may generally be neglected, so that H and U become practically equivalent.

Experimentally, the derivation of ΔH_1 and ΔS_1 rests on the relation between these quantities and the equilibrium partial vapour pressures of the components in the mixed phase. In the case when only one component (the low-molecular-weight liquid) has an appreciable vapour pressure the molar free energy of dilution is given by

$$\Delta G_1 = RT \ln (p/p_0), \qquad (7.4)$$

where p is the vapour pressure of the liquid component in equilibrium with the mixture (swollen rubber) at the temperature T and p_0 is its saturation vapour pressure. (This equation assumes that the vapour obeys the perfect gas laws; if this assumption is not fulfilled a correction must be introduced.) The heat of dilution is thermodynamically related to the temperature dependence of relative vapour pressure, thus

$$\Delta H_1 = \frac{\partial(\Delta G_1/T)}{\partial(1/T)} = - RT^2 \frac{\partial \ln (p/p_0)}{\partial T}. \qquad (7.5)$$

Eqns (7.4) and (7.5) enable ΔG_1 and ΔH_1 to be determined by experiment; the remaining quantity ΔS_1 is then given by the difference between these two quantities (eqn (7.2)).

An alternative method of obtaining ΔH_1 is by a direct calorimetric measurement of the heat of mixing. From the definition of heat content it follows that the change of heat content is equal to the heat absorbed on mixing of the two components at constant temperature. This method may be difficult to apply in the case of a rubber, on account of the slowness of the liquid absorption process, but where it can be applied it provides a useful check on the data obtained from vapour-pressure measurements.

Eqn (7.4) defines the relation between the free energy of dilution and the vapour pressure of the liquid component of the swollen rubber for any composition or degree of swelling. For the case when the polymer is in contact with the pure liquid the condition for equilibrium between the two phases, from eqns (7.1) and (7.4), becomes

$$RT \ln (p/p_0) = 0, \qquad (7.4a)$$

or $p/p_0 = 1$, indicating that the equilibrium vapour pressure of the liquid in the mixed phase is equal to the vapour pressure of the pure liquid. This conclusion, which may be regarded as self-evident, implies simply that the equilibrium degree of swelling will be the same whether the polymer is in direct contact with the pure liquid, or whether it is in indirect communication with the liquid via the saturated vapour.

It may be noted in passing that the condition defined by eqn (7.4a) need not necessarily correspond to a finite or limited degree of swelling. For an unvulcanized rubber in a good solvent the 'swelling' may proceed to an unlimited extent, the rubber and liquid being miscible or mutually soluble in all proportions. In such a case, eqn (7.4a) is satisfied only at infinite dilution (zero concentration) of the polymer. In a vulcanized rubber, on the other hand, the swelling is necessarily limited, since unlimited swelling could not take place without disruption of the chemically bonded network structure.

The free energy of dilution may be obtained not only from direct vapour-pressure measurements, but also, under suitable conditions, from any one of a number of properties which are thermodynamically related to the vapour pressure (*colligative* properties). Of these, the most important is the osmotic pressure Π. This is defined as the excess pressure which it is necessary to apply to the mixed phase in order that it shall be in equilibrium with the pure liquid. The appropriate relation is

$$\Pi = -(RT/V_1) \ln (p/p_0), \qquad (7.6)$$

where V_1 is the molar volume of the pure liquid. The quantity Π may also represent the *swelling pressure*, the only difference between these quantities being the purely practical one that for an osmotic pressure measurement a semi-permeable membrane is required in order to separate the two phases (which are both liquid), while in the measurement of swelling pressure (where the mixed phase is solid), the separating membrane becomes superfluous. Other solution properties of less general interest which are related to the vapour pressure are the elevation of the boiling point and the depression of the freezing point, with respect to the pure liquid.

7.3. Experimental data

As an example of the application of these thermodynamic principles, we may consider the experimental data obtained by Gee and his associates for the swelling (or solution) of unvulcanized natural rubber in benzene. Fig. 7.1 represents the relative vapour pressure p/p_0 of the solvent as a function of the volume fraction v_1 of

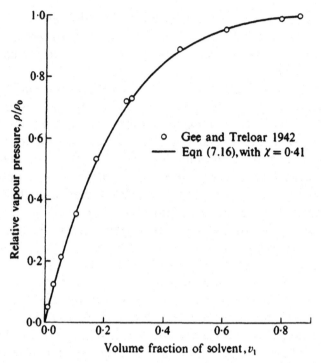

FIG. 7.1. Relative vapour pressure for rubber–benzene mixtures.

the solvent in the mixture. In this system the two components are miscible in all proportions, and the vapour pressure therefore differs from that of the pure liquid for all values of v_1, though for concentrations of solid below about 10 per cent ($v_1 > 0.9$) the difference is too small to be apparent in this figure. From these vapour pressure data the corresponding values of ΔG_1, the free energy of dilution, may be directly calculated, using eqn (7.4); the results so obtained are shown in Fig. 7.2. This quantity is always negative, with values ranging from $-\infty$ at $v_1 = 0$ to zero at $v_1 = 1$.

Heats of dilution ΔH_1, whether obtained from the temperature coefficient of relative vapour pressure or by direct calorimetry, cannot be measured with the same accuracy as the corresponding free energies. This is apparent from Fig. 7.3, which shows the values obtained from the temperature coefficients of vapour pressure and of osmotic pressure, using eqn (7.5). In this figure the quantity plotted is $\Delta H_1/v_2^2$, where $v_2 = 1 - v_1$. The experimental accuracy was obviously not sufficient to determine the form of the curve, but

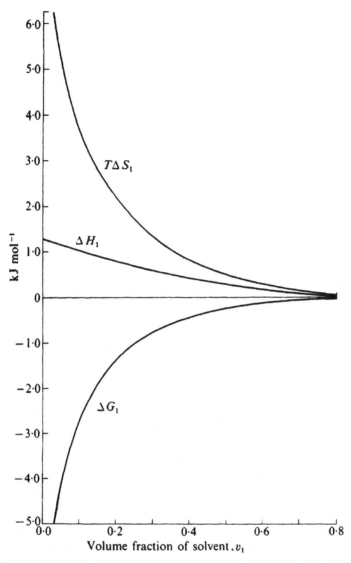

FIG. 7.2. Free energy, heat, and entropy of dilution for the system rubber–benzene, calculated from vapour pressure and osmotic data.

this was inferred from more accurate direct calorimetric measurements on a polyisoprene of low molecular weight (Gee and Orr 1946). The corresponding curve for ΔH_1 is included also in Fig. 7.2, from which it is seen that the maximum value of this quantity is 1300 J mol^{-1}, or about 17 J (4 cal) per gram of benzene.

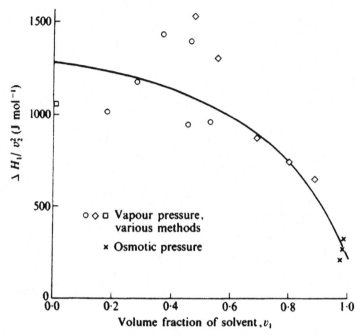

FIG. 7.3. The quantity $\Delta H_1/v_2^2$ for the system rubber–benzene (Gee and Orr 1946).

Finally, the entropy of dilution, obtained from ΔG_1 and ΔH_1 by the use of eqn (7.2), has the form shown in Fig. 7.2.

7.4. Significance of thermodynamic quantities

Important conclusions may be drawn immediately from the relative magnitudes of ΔG_1, ΔH_1, and $T\,\Delta S_1$ shown in Fig. 7.2. First, it is clear that the most important term is $T\,\Delta S_1$, which is always positive and relatively large. The heat term ΔH_1 is relatively small, and is in this case positive, corresponding to an absorption of heat on mixing. The positive sign means that this term acts in the sense of reducing the numerical value of ΔG_1, i.e. it tends to *oppose* the process of swelling or solution. Similar results are obtained with other rubbers in good solvents, though the sign of ΔH_1 may in some cases, e.g. rubber–chloroform, be negative. Our conclusion, therefore, is that the 'driving force' in the process of swelling or solution is the large increase in the associated entropy, the corresponding change in heat content or internal energy being relatively unimportant.

A process which would satisfy these thermodynamic conclusions is the process of diffusion or random intermixing of the two components. Such a process is accompanied by a large increase in the entropy of the system. In addition, if no strong intermolecular forces of a chemical or quasi-chemical character are brought into play, the change in heat content or internal energy will be relatively small. If this explanation is accepted, it follows that in all important respects the phenomenon of the swelling of a rubber in a typical solvent is analogous to the mixing of two mutually soluble low molecular weight liquids which do not interact chemically, e.g. benzene and petrol. A more detailed comparison shows that the values of ΔH_1 for typical rubber–liquid systems are in the same range as the values for simple liquids of comparable chemical constitution, as would be expected from the consideration that the intermolecular force fields, on which the value of ΔH_1 depends, are not markedly affected by the *length* of the molecules concerned, i.e. by their polymeric nature. The term ΔS_1, however, as is shown in more detail in the following section, is associated directly with the *configurational arrangements* of the molecules and is therefore greatly affected by their length. The entropy of dilution, though not different in sign, is therefore very different in magnitude in the case of a polymer than in the case of liquid–liquid mixtures.

It is apparent, therefore, that the facility for swelling does not imply any specific chemical attraction between the rubber and liquid molecules. It is a purely physical mixing or interdiffusion process in which the two components may be regarded as chemically neutral.

There is a close thermodynamic analogy between the elastic properties of a rubber and the phenomenon of swelling. Both are manifestations of the configurational entropy of the system of long-chain molecules. The increase in entropy on elastic retraction or on mixing are both in harmony with the principle that any spontaneous process is accompanied by an increase in the entropy of the system. In both cases also this primary effect is accompanied by a secondary and essentially irrelevant change in internal energy arising from intermolecular forces (cf. Chapter 2).

The above characteristic features of the phenomenon of swelling in rubbers are in marked contrast to the typical characteristics of water-swelling substances. In these, the quantity ΔH_1 is generally *negative* and large (i.e. heat is *evolved* on absorption of water) while the entropy is *negative* and smaller (cf. Katz 1933). In such systems, among which cellulose (cotton) may be taken as typical, the driving force in the process of swelling is the chemical attraction between water and polymer molecules (via hydrogen bonds), which is effective in spite of the reduction in entropy.

7.5. Statistical treatment of swelling

The essential problem in the statistical treatment of swelling is to determine the increase of entropy which accompanies the mixing of polymer and liquid molecules. This increase in entropy arises from the greater probability of the mixed state compared with the unmixed state, and may be calculated in terms of the number of configurations available to the system at any given composition.

Comparatively simple methods for the calculation of the configurational entropy of mixing have been worked out by Flory (1942) and by Huggins (1942). In Flory's model the liquid and polymer molecules are considered for convenience of calculation to be arranged on a three-dimensional lattice of sites such that each site may be occupied either by a liquid molecule or by a single segment of a polymer chain. While a liquid molecule is free to occupy *any* vacant site, the successive segments of a particular polymer molecule are, of course, restricted to adjacent sites, as illustrated in the accompanying two-dimensional diagram (Fig. 7.4). If n_0 is the total number of sites and N the number of polymer

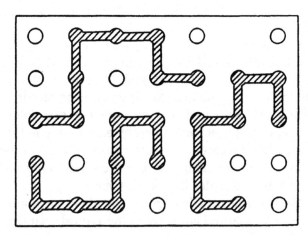

FIG. 7.4. Schematic representation of lattice model. The circles represent solvent molecules.

molecules, each consisting of x segments, the number of liquid molecules is $n_0 - xN$. The problem is to calculate the total number of distinguishable configurations or conformations of the N polymer molecules and $n_0 - xN$ liquid molecules on the lattice. For this calculation Flory considers the polymer molecules to be laid down successively, segment by segment, on the lattice. When these have been dealt with the liquid molecules are allowed to fill the remaining vacant sites.

Let us consider the state existing after N_i polymer molecules have been introduced, and let us calculate the number of ways in which the next polymer molecule may be laid down. Its first segment may be placed on any one of the $n_0 - xN_i$ vacant sites. The second segment may be placed on any one of the Z nearest-neighbour sites to that occupied by the first segment that is not already occupied by a polymer segment. The fraction of unoccupied sites on the whole lattice being $(n_0 - xN_i)/n_0$, the number of sites available to the second segment is taken as $Z(n_0 - xN_i)/n_0$, i.e. proportional to the average concentration of unoccupied sites. Similarly the number of sites available to the third segment is

$$\alpha = (Z - 1)(n_0 - xN_i)/n_0, \qquad (7.7)$$

since one of the Z nearest-neighbour sites is already occupied by the second segment. This same expression is also used to represent the number of sites available for all later segments; this means that the complication due to the probability of occupancy of a potentially vacant site by earlier segments of the same chain is ignored. The total number of conformations for the $(i + 1)$th chain is thus

$$\nu_{i+1} = \tfrac{1}{2}(n_0 - xN_i)\left(\frac{Z}{Z-1}\right)\alpha^{x-1} \qquad (7.8)$$

the factor $\tfrac{1}{2}$ being introduced because either end of the chain may be taken as the starting-point. The total number of distinguishable conformations for the assembly of N polymer molecules is then

$$W = \frac{1}{N!} \prod_{i=1}^{N} \nu_i, \qquad (7.9)$$

where ν_i is obtainable from (7.8).

The configurational entropy ΔS is obtained from (7.9) by the use of Boltzmann's relation $S = k \ln W$. This gives, after further reduction,

$$\Delta S = -k\left(n \ln \frac{n}{n+xN} + N \ln \frac{xN}{n+xN}\right) + k(x-1)N[\ln (Z-1)-1] -$$
$$- kN \ln 2, \tag{7.10}$$

where $n = n_0 - xN$ is the number of liquid molecules. Subtracting from this expression for ΔS the configurational entropy corresponding to the original unswollen polymer, obtained by putting $n = 0$ in eqn (7.10), the entropy ΔS_m associated with the mixing process is obtained in the form

$$\Delta S_m = -k\left(n \ln \frac{n}{n+xN} + N \ln \frac{xN}{n+xN}\right) \tag{7.11}$$

or

$$\Delta S_m = -k(n \ln v_1 + N \ln v_2), \tag{7.11a}$$

where v_1 and v_2 are the respective volume fractions of liquid and polymer in the mixture. Note that the configurations of the liquid molecules are not directly involved in the above calculation. This is because the sites remaining vacant after the introduction of the polymer molecules can be filled in one way only.

The entropy of dilution ΔS_1 with respect to the liquid component is obtained by differentiation of (7.11) with respect to liquid content (n). In terms of molar quantities the result finally obtained (on writing $1 - v_2$ for v_1) is

$$\Delta S_1 = -R\{\ln (1-v_2) + (1-1/x)v_2\}. \tag{7.12}$$

Free energy of dilution

To obtain the free energy of dilution ΔG_1, it is necessary to introduce an expression for the heat of dilution ΔH_1. This is done by Flory on a semi-empirical basis, using the formula

$$\Delta H_1 = \alpha v_2^2. \tag{7.13}$$

This formula, which has some theoretical justification, has been found to apply to a large number of simple liquid mixtures. Introduction of eqns (7.12) and (7.13) into (7.2) leads to the following expression for the Gibbs free energy of dilution

$$\Delta G_1 = RT\{\ln (1-v_2) + (1-1/x)v_2 + (\alpha/RT)v_2^2\}. \tag{7.14}$$

The alternative treatment of the problem by Huggins (1942) follows the same general lines as that of Flory, but includes a more precise analysis of the number of sites available to segments of the polymer molecule subsequent to the third. His expression for the entropy of dilution is similar to (7.12), but contains an additional term in v_2^2. On inclusion of a heat of dilution of the form (7.13), exactly as in Flory's method, the resultant expression for the free energy of dilution becomes

$$\Delta G_1 = RT\{\ln(1 - v_2) + (1 - 1/x)v_2 + \chi v_2^2\}, \qquad (7.15)$$

in which χ is a parameter which includes a component χ_0 due to entropy in addition to the purely energetic contribution ΔH_1, i.e.

$$\chi = \chi_0 + \alpha/RT, \qquad (7.15a)$$

where χ_0 and α are constants. At any given temperature, however, eqns (7.14) and (7.15) are formally identical. Furthermore, if the number of segments in the polymer chain is sufficiently large, so that $1/x$ is very small, either of these equations reduces to the form

$$\Delta G_1 = RT\{\ln(1 - v_2) + v_2 + \chi v_2^2\}, \qquad (7.16)$$

in which χ may be regarded as a constant. This is the form usually referred to as the Flory–Huggins equation.

It is a remarkable feature of this result that it includes only a single adjustable parameter χ which is dependent on the special properties of the particular polymer–liquid combination represented. This suggests that many of the restrictions introduced by the somewhat artificial lattice model are formally unnecessary and irrelevant to the essential physical problem, and that the resultant equation should have a degree of generality exceeding that which would be strictly justified on the basis of the original model.

7.6. Comparison with experiment

The application of the Flory–Huggins eqn (7.16) to the vapour-pressure data of Gee and Treloar (1942) for the rubber–benzene system is shown in Fig. 7.1, where the continuous curve represents the theoretical relation, with $\chi = 0{\cdot}41$. The agreement over the whole range of composition is seen to be very good. However, the analysis of the free energy into a heat and an entropy term yields a less satisfactory result. According to eqn (7.13) the quantity $\Delta H_1/v_2^2$ should be constant, for variations of composition v_2. Fig. 7.3 shows that this expectation is not fulfilled. It must therefore be concluded that the deviations from the theoretical form of ΔH_1 are compensated by deviations in ΔS_1 in such a way that the resultant free energy is unaffected. From the present point of view, however, these divergencies are not serious, since our primary concern is with the form of the free energy ΔG_1. This is adequately represented by the Flory–Huggins equation, with χ treated as an adjustable parameter. Typical values of this parameter, taken from a review article by Sheehan and Bisio (1966) are given in Table 7.1.

The significance of the constant χ is further examined in § 7.9 below.

7.7. The swelling of cross-linked polymers

An examination of eqn (7.16) shows that if $\chi < \frac{1}{2}$, ΔG_1 is negative for all values of v_2, which means that the polymer and liquid are miscible in all proportions, i.e. the polymer is soluble. On the other hand, if $\chi > \frac{1}{2}$, there is a particular value of v_2 for which $\Delta G_1 = 0$; this represents the condition for limited or equilibrium swelling.

These considerations apply only when the polymer is not cross-linked. In a cross-linked polymer, as we have already noted, the presence of an interconnected network precludes the possibility of solution. For such a material the Flory–Huggins equation is not in itself sufficient to give the free energy of dilution; it is necessary to take into account also the configurational entropy of the network. The total free energy of dilution must therefore be expressed as the sum of two terms, i.e.

$$\Delta G_1 = \Delta G_{1m} + \Delta G_{1e}, \tag{7.17}$$

where ΔG_{1m} represents the free energy of dilution for the polymer in the state prior to cross-linking and ΔG_{1e} corresponds to the change of free energy (per mole of liquid absorbed) due to the associated elastic expansion of the network. This second term is obtained from the Gaussian network theory, as represented by the expression (4.9a) for the elastically stored free energy. For an isotropic expansion of the network we have $\lambda_1 = \lambda_2 = \lambda_3 = v_2^{-\frac{1}{3}}$ (where $1/v_2$ is the volume swelling ratio); the change in network free energy on swelling is therefore

$$W = \frac{\rho RT}{2M_c}(\lambda_1^2 + \lambda_2^2 + \lambda_3^2 - 3) = \frac{3\rho RT}{2M_c}(\lambda_2^{-\frac{2}{3}} - 1). \tag{7.18}$$

TABLE 7.1

Values of Flory–Huggins parameter χ. (The third decimal figure is probably not significant)

Liquid	Natural rubber	Poly-chloroprene (Neoprene)	Butyl rubber	Butadiene-styrene rubber (71·5 per cent styrene)	Butadiene-acrylonitrile rubber (18 per cent acrylonitrile)	Silicone rubber (dimethyl siloxane)
Benzene	0·421	0·263	0·578	0·398		0·52
Toluene	0·393		0·557			0·465
Hexane	0·480	0·891	0·516	0·656	0·990	0·40
Decane	0·444	1·147	0·519	0·671	1·175	
Dichloromethane	0·494	0·533	0·579	0·474	0·394	
Carbon tetrachloride	0·307		0·466	0·362	0·478	0·45
n-Propyl acetate	0·649					
Methyl ethyl ketone	0·856					
Acetone	1·36					

The corresponding term in the molar free energy of *dilution* is $\partial W/\partial n_1$, where n_1 is the number of moles of liquid in the swollen polymer. Putting

$$1/v_2 = 1 + n_1 V_1 \qquad (7.19)$$

(where V_1 is the molar volume of the swelling liquid) into eqn (7.18) and differentiating, we obtain

$$\Delta G_{1e} = \frac{\rho RT}{M_c} V_1 v_2^{\frac{1}{3}}. \qquad (7.20)$$

Introducing this into eqn (7.17) together with the Flory–Huggins expression for ΔG_{1m} we obtain the total free energy of dilution in the form

$$\Delta G_1 = RT\left\{\ln(1-v_2) + v_2 + \chi v_2^2 + \frac{\rho V_1}{M_c} v_2^{\frac{1}{3}}\right\}. \qquad (7.21)$$

The condition for equilibrium swelling, $\Delta G_1 = 0$, is therefore

$$\ln(1-v_2) + v_2 + \chi v_2^2 + \frac{\rho V_1}{M_c} v_2^{\frac{1}{3}} = 0. \qquad (7.22)$$

The equilibrium degree of swelling is represented by the value of v_2 which satisfies this equation.

If the modified formula of Flory (eqn (4.28)) were used in place of eqn (4.20) to represent the network entropy, eqn (7.18) would be replaced by

$$W = \frac{\rho RT}{2M_c}(3v_2^{-\frac{2}{3}} - \ln v_2^{-1} - 3), \qquad (7.18\,a)$$

and the equation for equilibrium swelling would become

$$\ln(1-v_2) + v_2 + \chi v_2^2 + \frac{\rho V_1}{M_c}\left(v_2^{\frac{1}{3}} - v_2/2\right) = 0. \qquad (7.22a)$$

7.8. Relation between swelling and modulus

Eqn (7.22) expresses the relation between the equilibrium swelling and the degree of cross-linking, represented in terms of the molecular weight M_c of the network chains. The latter quantity, in turn, is directly related to the elastic (shear) modulus G, given by the elementary Gaussian network theory in the form (4.9c), i.e.

$$G = NkT = \rho RT/M_c.$$

For a given polymer subjected to varying amounts of cross-linking there is thus a unique relationship between the equilibrium degree of swelling (in a given liquid) and the modulus.

This relationship between swelling and modulus has been the subject of a number of experimental studies, starting with the early work of Flory (1944, 1946) and Gee (1946b). By expanding the logarithm in eqn (7.22) and neglecting powers of v_2 higher than the second, Flory showed that for high degrees of swelling this relationship could be reduced to the approximate form

$$(\chi - \tfrac{1}{2})v_2^2 + (\rho V_1/M_c)v_2^{\frac{1}{3}} \simeq 0$$

or

$$(\rho V_1/M_c) \simeq (\tfrac{1}{2} - \chi)v_2^{\frac{5}{3}}. \qquad (7.22b)$$

Using a series of differently cross-linked butyl rubbers, Flory (1944) presented his results in the form of a double logarithmic plot of the force per unit unstrained area at a given strain (which is proportional to $\rho RT/M_c$) against the equilibrium degree of swelling $(1/v_2)$ in cyclohexane. The data (Fig. 7.5) fell on a straight line of slope $-\tfrac{5}{3}$, in agreement with eqn (7.22b). Gee, working with natural rubber, calculated the values of modulus from measurements of the

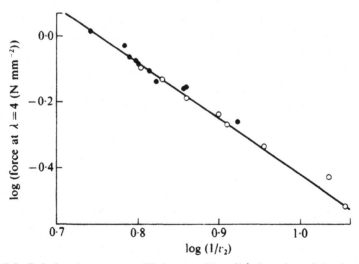

FIG. 7.5. Relation between equilibrium swelling $(1/v_2)$ and modulus for butyl rubbers with various degrees of cross-linking (From Flory 1946).

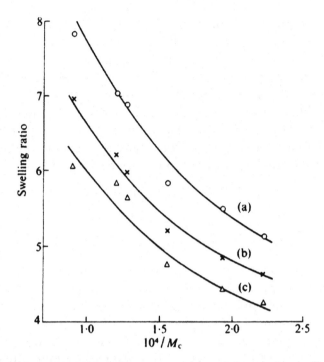

FIG. 7.6. Relation between equilibrium swelling and degree of cross-linking for natural rubber in (a) carbon tetrachloride (b) carbon disulphide, and (c) benzene. Curves from eqn (7.22) (Gee 1946*b*).

force–deformation relations in the *swollen* state, thereby circumventing the difficulties associated with the deviations from the statistical theory (Chapter 5). By an appropriate choice of χ, reasonably close agreement was obtained with the theory, particularly in the case of good swelling agents (Fig. 7.6). Moreover, as will be seen from Table 7.2, these values of χ were in close agreement with the values obtained from direct vapour-pressure measurements.

    More extensive studies of the relation between swelling and modulus have been carried out by Mullins, who worked mainly with peroxide-vulcanized rubbers, but included some conventional sulphur vulcanizates for comparison. Cross-linking by peroxides has the advantage of producing direct C—C cross-linkages between chains, without the introduction of any non-rubber constituents into the network structure. To overcome the difficulties arising from deviations from the form of force–extension curve predicted by the statistical theory (as a result of which the value of modulus depends

TABLE 7.2

Values of χ from swelling and from vapour-pressure measurements
(Gee 1946*b*)

Liquid	From swelling	From vapour pressure
Carbon tetrachloride	0·290	0·28
Chloroform	0·340	0·37
Carbon disulphide	0·425	0·49
Benzene	0·395	0·41
Toluene	0·360	0·43–0·44
Petroleum (light)	0·540	0·43

on the magnitude of the strain at which the measurements are made) Mullins (1956) worked with rubbers in the highly swollen state, corresponding to $v_2 < 0\cdot25$, where these deviations are eliminated, i.e. where $C_2 = 0$ (cf. Chapter 5). In cases where a sufficiently high degree of swelling was not obtainable values obtained by extrapolation from lower degrees of swelling were employed. An empirical correction for 'loose ends', based on data obtained for rubbers of different molecular weight M before cross-linking, was applied; this is discussed in Chapter 8, in which the methods employed are presented in detail. The values of modulus so obtained were converted to equivalent values of modulus for the unswollen rubber by multiplying by the factor $v_2^{-\frac{1}{3}}$, in accordance with the statistical theory (eqn (4.26)); the quantity finally obtained was designated by the symbol $2C_1$. Fig. 7.7 shows the relation between the constant $2C_1$ obtained in this way and the value of v_2 corresponding to the equilibrium degree of swelling in *n*-decane. The results are in close agreement with the Flory–Huggins relation (eqn (7.22)), with a value of χ of $0\cdot41_3$. Equally good agreement was obtained with the modified formula (7.22a); in this case the value of χ required to fit the data was $0\cdot45_5$.

In a subsequent paper Mullins (1959*b*) noted that at high degrees of swelling the effects of the limited extensibility of the network (non-Gaussian effects) could produce a significant increase in the modulus even in the region of small extensions. He concluded that for this reason values of C_1 obtained from measurements on the unswollen rubbers should be preferred to those derived from measurements on swollen rubbers. The values obtained on this

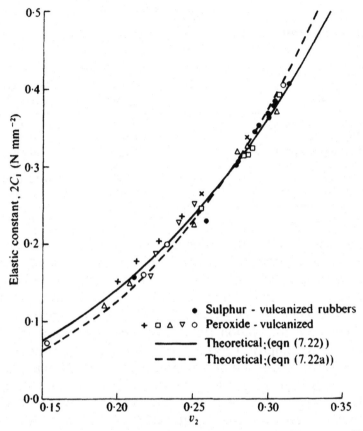

FIG. 7.7. Relation between elastic constant $2C_1$ and v_2 for equilibrium swelling in n-decane. (From Mullins 1956.)

basis were some 10 per cent lower than the previously obtained values. At the same time minor improvements were introduced into the calculation of the molecular weights prior to cross-linking. The final results obtained using these modifications showed a similar agreement with the Flory–Huggins relation (7.22) to that obtained originally, but necessitated a small adjustment in the value of χ, from 0·41 to 0·42.

It is necessary to emphasize that the values of $2C_1$ used in these experiments do not represent the shear modulus of the unswollen rubber, which for small strains is given by $2(C_1 + C_2)$. The use of $2C_1$ is based on the assumption that it is this constant alone which is directly related to the network properties, and which should therefore be identified with the value of the shear modulus given by the statistical theory, namely, $\rho RT/M_c$. Possible objections to this assumption are discussed in detail in Chapter 8.

7.9. The cohesive-energy density

Up to the present we have considered the specific differences in swelling behaviour among the various polymer–liquid systems to be capable of representation by means of the empirical parameter χ, which we saw was related primarily, though not entirely, to the heat of mixing or energy associated with the interactions between polymer and liquid molecules. In this section we examine in more detail the nature of this interaction and consider alternative methods of representing its numerical magnitude.

The most important consideration in this connection is that the energetic interactions between polymer and solvent molecules, as previously noted, are not specifically related to the polymeric nature of the polymer component but arise from the local fields of force between neighbouring atoms; they are therefore likely to resemble very closely the energetic interactions between corresponding pairs of low-molecular-weight liquids of comparable chemical constitution. Bearing this in mind, Gee (1942, 1943) attempted to apply the concept of the *cohesive-energy density* to obtain a semi-quantitative understanding of the role of intermolecular interactions in the determination of swelling properties. The cohesive-energy density (c.e.d.) is defined as the energy required to separate all the molecules in a given material from one another; its value (per mole) is equal to $(L - RT)/V$, where L and V are respectively the molar latent heat of evaporation and molar volume. For many non-polar liquid mixtures it has been found that the heat of dilution ΔH_1 (for component 1) may be represented by the expression

$$\Delta H_1 = kV_1(e_1^{\frac{1}{2}} - e_2^{\frac{1}{2}})^2 v_2^2, \tag{7.23}$$

in which e_1 and e_2 are the respective c.e.d. values for the two liquids, v_2 is the volume fraction of component 2, and k is a numerical factor. According to this equation, the heat of dilution is always positive, and passes through a minimum (zero) when $e_1 = e_2$.

Adapting this idea to the case of a polymer–liquid mixture, Gee suggested that, for a given polymer swollen in a variety of liquids, the swelling should be a maximum when ΔH_1 is a minimum, i.e. for that liquid whose c.e.d. is equal to that of the polymer. He found that a plot of the volume swelling ratio for a given vulcanized rubber against the c.e.d. of the swelling liquid yielded a curve with a pronounced maximum; the position of this maximum was therefore taken to be a measure of e_2, the c.e.d. of the polymer, which of course cannot be measured directly. In this way he derived the value $e_2 = 63 \cdot 7 \text{ cal cm}^{-3}$ (266 J cm^{-3}) for natural rubber. A plot of swelling against $V_1^{\frac{1}{3}}(e_1^{\frac{1}{2}} - e_2^{\frac{1}{2}})$, in accordance with eqn (7.23), then yielded a curve having the form shown in Fig. 7.8.

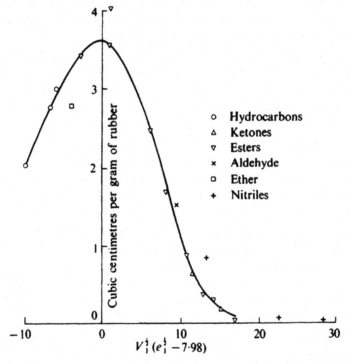

FIG. 7.8. Relation of swelling to cohesive-energy density. (Gee 1943.)

The solubility parameter

Equations such as (7.23) may be presented rather more conveniently in terms of a parameter known as the 'solubility parameter', which is defined as the square root of the c.e.d., and is denoted by the symbol δ. Numerical values of this parameter for a large number of polymers, obtained by a method similar to that of Gee, and also for solvents of various types, have been collected by

Sheehan and Bisio (1966); a selection from these figures is given in Table 7.3. The values of δ for the various solvents change systematically with their chemical constitution and tend to increase with increasing polarity. The best solvents (or swelling agents) for a

TABLE 7.3

Values of solubility parameter δ *(cal cm^{-3})$^{\frac{1}{2}}$*

Polymers	
Natural rubber	8·10
Butyl rubber	7·84
Polybutadiene	8·44
Neoprene (polychloroprene)	8·85
Butadiene–styrene (28·5 per cent styrene)	8·33
Butadiene–acrylonitrile (18 per cent acrylonitrile)	8·70
Butadiene–acrylonitrile (39 per cent acrylonitrile)	10·30
Polyethylene	7·94
Solvents	
Hexane	7·33
Decane	7·77
Cyclohexane	8·25
Benzene	9·22
Toluene	8·97
Chloroform	9·30
Carbon tetrachloride	8·63
Acetone	9·74
Ethanol	12·97
Methanol	14·52

given polymer are those whose δ-values are nearest to that of the polymer, and which therefore are most closely related in chemical structure to the polymer. Thus hydrocarbon rubbers (natural rubber, butyl rubber, polybutadiene) are readily soluble in hydrocarbon solvents (petrol, benzene, etc.), but insoluble in polar liquids such as acetone and alcohol. For the more polar butadiene–acrylonitrile rubbers the value of δ increases with increasing acrylonitrile content, with a consequent increasing resistance to absorption of hydrocarbon solvents such as petrol and lubricating oils.

In view of the association of both χ and cohesive-energy density with the heat of dilution, represented by the respective equations (7.15a) and (7.23), a close relationship between the two parameters χ and δ is also to be expected. The formulation of this relationship, however, is complicated by the presence of both entropic (χ_s) and energetic (χ_H) contributions to the observed χ-value, and it is only the χ_H component which is relevant in this connection. On the basis of experimental data for non-polar polymer–solvent systems Blanks and Prausnitz (1964) estimated the mean numerical value of χ_s at 0·34. This value may be used, in conjunction with an equation given by Shvarts (1958), namely,

$$\delta_1 = \delta_2 \pm \{(RT/V_1)(\chi - \chi_s)\}^{\frac{1}{2}}, \qquad (7.24)$$

which is substantially the same as (7.23), to relate the values of δ and χ for a variety of liquids, and has proved moderately successful in accounting for the experimental data (Sheehan and Bisio 1966).

7.10. The dependence of swelling on strain

General equations

So far we have been concerned with the question of the equilibrium swelling of a cross-linked rubber in the stress-free state in contact with a liquid, and have seen that this can be satisfactorily accounted for in terms of the Flory–Huggins theory. We now turn to the rather more complicated problem of determining the equilibrium degree of swelling in the presence of a stress or mechanical restraint. This problem was first solved by Flory and Rehner (1944) and also by Gee (1946b) for the case of a simple tensile stress. Rather surprisingly (at first sight) it was found that the effect of the tensile stress is to *increase* the amount of the swelling, compared with that for the stress-free rubber. In the case of a good swelling agent the effect is quite large and may readily be confirmed experimentally.

We shall begin by considering the theory for the general case, corresponding to a pure homogeneous strain of any type, and will then apply the result obtained to more specific types of strain. This problem has been worked out by the writer (Treloar 1950a). As in the previous treatment (§ 7.6) the total change of free energy is represented as the sum of two terms, one of which corresponds to the mixing of polymer and liquid molecules, and the other to the free energy (entropy) of network deformation. In addition, however, it is necessary to take into account the work done by the forces which have to be applied to maintain the specified state of strain.

Let us consider a specimen of the cross-linked material, originally in the form of a unit cube, which is in contact with the liquid, and is constrained by normal forces applied to its faces to the form of a rectangular block having dimensions l_1, l_2, and l_3 (Fig. 7.9). The

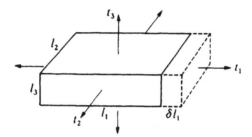

FIG. 7.9. Equilibrium of swollen rubber under stress.

edges of the block define the principal axes of the strain, while the applied forces determine the principal stresses t_1, t_2, and t_3. If the volumes of the polymer and liquid are assumed to be additive we shall have

$$l_1 \, l_2 \, l_3 = 1/v_2 = 1 + n_1 V_1, \qquad (7.25)$$

where n_1 is the number of moles of liquid in the swollen polymer and V_1 is the molar volume of the liquid, v_2 being the volume fraction of polymer, as previously. According to our initial assumption we may write

$$\Delta G = \Delta G_m + \Delta G_e, \qquad (7.26)$$

where ΔG is the total change in the Gibbs free energy of the system (polymer plus liquid) on passing from the *unstrained unswollen* state to the *strained swollen* state and ΔG_m and ΔG_e are the terms representing the corresponding free energies of mixing and of network deformation.

Consider now the process of absorption of a further small quantity δn_1 moles of liquid, under the condition that l_2 and l_3 are held constant, while l_1 increases by the amount δl_1. We note that the stress t_1 acts on a surface area $l_2 l_3$; the corresponding force on this surface is therefore $t_1 l_2 l_3$. Since no work is done by t_2 and t_3, the total work δW done by the external forces is simply

$$\delta W = t_1 l_2 l_3 \delta l_1 = t_1 V_1 \, \delta n_1 \qquad (7.27)$$

from (7.25) above. Under constant temperature conditions, which will be assumed throughout, the condition governing the equilibrium of a system which is subjected to the action of external forces is that the change in the Helmholtz free energy ΔA in a small displacement from equilibrium shall be equal to the work done by the applied forces, i.e.,

$$\delta \Delta A = \delta W. \tag{7.28}$$

From the definition of *Gibbs* free energy the relation between ΔG and ΔA, under conditions of constant pressure p and temperature T, is

$$\Delta G = \Delta A + p\Delta V. \tag{7.29}$$

In the present case the net volume change (of liquid plus polymer) is assumed to be zero, and we may therefore put $\Delta G = \Delta A$. Hence, from (7.27) and (7.28),

$$\delta \Delta G = \delta W = t_1 V_1 \, \delta n_1, \tag{7.30}$$

or, in terms of partial differentials,

$$\left(\frac{\partial \Delta G}{\partial n_1}\right)_{l_2 l_3} = t_1 V_1 \tag{7.31}$$

Further, from (7.26) we have

$$\left(\frac{\partial \Delta G}{\partial n_1}\right)_{l_2 l_3} = \frac{\partial \Delta G_m}{\partial n_1} + \left(\frac{\partial \Delta G_e}{\partial n_1}\right)_{l_2 l_3}, \tag{7.32}$$

where the first term on the right is the free energy of dilution for the polymer in the uncrossed-linked state. For this we shall assume the Flory–Huggins relation (7.16), which in the present notation becomes

$$\frac{\partial \Delta G_m}{\partial n_1} = RT\{\ln (1 - v_2) + v_2 + \chi v_2^2\}. \tag{7.33}$$

To obtain the second term we make use of the expression (4.9a) for the elastic free energy of the swollen network referred to the unswollen state; this gives

$$\Delta G_e = W = \frac{\rho RT}{M_c}(l_1^2 + l_2^2 + l_3^2 - 3). \tag{7.34}$$

We thus obtain, with the aid of eqn (7.25)

$$\left(\frac{\partial \Delta G_e}{\partial n_1}\right)_{l_2 l_3} = \left(\frac{\partial \Delta G_e}{\partial l_1}\right)\left(\frac{\partial l_1}{\partial n_1}\right)_{l_2 l_3} = \frac{\rho RT}{M_c} l_1 \frac{V_1}{l_2 l_3} = \frac{\rho V_1 RT}{M_c} v_2 l_1^2.$$

(7.35)

Combining (7.33) and (7.35) according to (7.32), to obtain the total free energy of dilution, and introducing the equilibrium condition (7.31), we arrive at the final result

$$t_1 = \frac{RT}{V_1}\left\{\ln(1-v_2) + v_2 + \chi v_2^2 + \frac{\rho V_1}{M_c} v_2 l_1^2\right\}.$$

(7.36)

This equation expresses the relation between the swelling ratio $(1/v_2)$, the dimension l_1, and the stress t_1. Corresponding relations may be written down for t_2 and t_3 in terms of l_2 and l_3, respectively. These equations together represent the solution to the problem.

Stress–strain relations for swollen rubber

From the three equations of the type (7.36) we obtain by subtraction three general stress–strain relations for the swollen rubber in the form

$$t_1 - t_2 = \frac{\rho RT}{M_c} v_2(l_1^2 - l_2^2) = \frac{\rho RT}{M_c} v_2^{\frac{1}{3}}(\lambda_1^2 - \lambda_2^2),$$

(7.37)

in which λ_1, λ_2 and λ_3 are the principal extension ratios referred to the *swollen* unstrained state. These equations have already been derived for a swollen rubber (p. 70); they do not involve the free energy of mixing, and are valid whether or not the rubber is in equilibrium with the liquid. The condition for equilibrium with respect to the liquid content determines the value of v_2, but not the mechanical properties of the rubber at any particular value of v_2. If v_2 is considered to be fixed, the swollen rubber may be regarded as incompressible, in which case only the *differences* of principal stresses are determinate, just as in the case of an unswollen rubber. However, if *in addition* we introduce the condition that the swollen rubber shall be in equilibrium with the swelling liquid, the further relationship (7.36) is introduced, whereby the values of the principal stresses become uniquely determined. Of the seven quantities l_1, l_2, l_3, t_1, t_2, t_3, and v_2, only three can be independently chosen; the remaining four are then determined by three equations of the type (7.36) together with eqn (7.25). Thus, for example, if the dimensions l_1, l_2, and l_3 are chosen as independent variables, the value of v_2 follows immediately from eqn (7.25), and the principal stresses are obtainable from eqn (7.36) together with the corresponding equations for t_2 and t_3.

The physical significance of this result is that whereas in an incompressible rubber (i.e. whether unswollen or swollen to a specified or fixed extent) the volume, and hence the state of strain, is unaffected by the superposition of an arbitrary hydrostatic pressure; in a *compressible* rubber, or in a rubber considered to be in equilibrium with respect to the swelling liquid, this is no longer true, since any such superimposed hydrostatic pressure will reduce the volume or liquid content (see below). In this respect, therefore, a swollen rubber in continuous equilibrium with a surrounding liquid may be regarded, from the purely formal standpoint, as having mechanical properties equivalent to those of a *compressible* material.

The application of the above analysis to particular cases of practical interest will now be considered.

The special case of hydrostatic pressure

In the case of a simple hydrostatic pressure p applied to the polymer (but not to the liquid), we have

$$t_1 = t_2 = t_3 = -p; \qquad l_1 = l_2 = l_3 = v_2^{-\frac{1}{3}}.$$

Hence for equilibrium we have, from eqn (7.36),

$$\frac{RT}{V_1}\left\{\ln(1-v_2) + v_2 - \chi v_2^2 + \frac{\rho V_1}{M_c}v_2^{\frac{1}{3}}\right\} = -p. \qquad (7.38)$$

The pressure p is equivalent to the swelling pressure (p. 131). For the particular case when $p = 0$, corresponding to free swelling, eqn (7.38) reduces to the form (7.22) given previously.

The special case of simple elongation (or uniaxial compression)

For a tensile stress t_1 acting in the direction l_1 we have

$$t_2 = t_3 = 0; \qquad l_2^2 = l_3^2 = 1/l_1 v_2.$$

Using the expression for t_2 corresponding to eqn (7.36) the equilibrium condition becomes

$$t_2 = \frac{RT}{V_1}\left\{\ln(1-v_2)+v_2+\chi v_2^2+\frac{\rho V_1}{M_c l_1}\right\} = 0. \qquad (7.39)$$

This equation may be solved for v_2, for a specified value of l_1, the length in the direction of the applied force. If l_1 is greater than $v_2^{-\frac{1}{3}}$ this force is tensile, while if l_1 is less than $v_2^{-\frac{1}{3}}$ it is compressive. For $l_1 = v_2^{-\frac{1}{3}}$ the tensile stress is zero, corresponding to free swelling.

The special case of equi-biaxial (two-dimensional) extension

This case is defined by the relations

$$t_2 = t_3, \qquad t_1 = 0; \qquad l_2^2 = l_3^2 = 1/l_1 v_2.$$

Hence

$$t_1 = \frac{RT}{V_1}\left\{\ln(1-v_2)+v_2+\chi v_2^2+\frac{\rho V_1}{M_c v_2 l_2^4}\right\} = 0. \qquad (7.40)$$

In this expression l_2 is the stretch ratio in the plane of the stretched sheet, referred to the unswollen unstrained dimensions.

7.11. Experiments on swelling of strained rubber

Experiments on simple extension were carried out originally by Flory and Rehner (1944), using butyl rubber swollen in xylene, and by Gee (1946b) using natural rubber in a variety of swelling agents. The theoretical relation between the extension and the degree of swelling was satisfactorily confirmed, except in the case of poor swelling agents, for which quantitative agreement was not obtained.

In a more extensive study the author (Treloar 1950b) examined natural rubber in simple extension, uniaxial compression and equi-biaxial extension, using both benzene and heptane as swelling liquids. The results are reproduced in Figs 7.10, 7.11, and 7.12. In the case of benzene, the value of χ was taken from published work on vapour pressures, and the value of M_c was then calculated from the equilibrium swelling in the unstrained state using eqn (7.22). For the swelling of the same rubbers in heptane, for which the value of χ was not available at the time, the same value of M_c was used, and χ was calculated from the free-swelling data. In either case the theoretical curve was therefore adjusted to fit the swelling data at one point only, i.e. for free swelling, but the dependence of

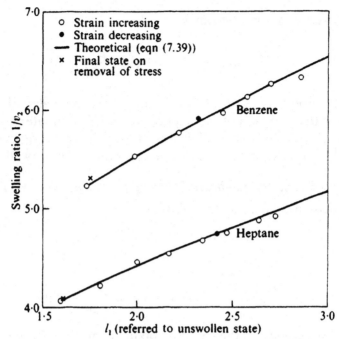

FIG. 7.10. Dependence of swelling on strain. Simple extension.

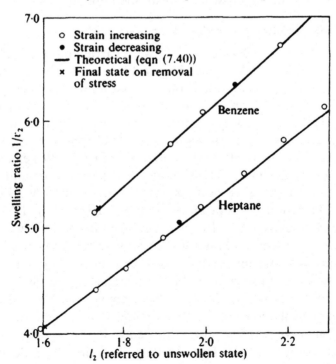

FIG. 7.11. Dependence of swelling on strain. Equi-biaxial extension.

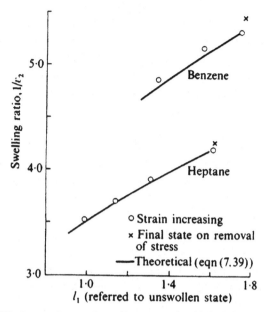

FIG. 7.12. Dependence of swelling on strain. Uniaxial compression.

swelling on strain is then theoretically determined, and cannot be further adjusted. The degree of correspondence between the theoretical and experimental curves therefore provides a critical test of the theory.

Origin of dependence of swelling on strain

At first sight the *increase* of swelling with extension may seem somewhat surprising. It may, however, be understood in principle in quite general terms. The quantity which determines the *direction* of the change of swelling is the hydrostatic component of the applied stress. For a simple hydrostatic pressure this leads directly to a reduction of swelling with increasing pressure. A uniaxial compressive stress t is equivalent to a hydrostatic pressure of magnitude $t/3$ together with two shear stresses of the same numerical value. It may be assumed (to a first approximation) that the shear stresses (for which the hydrostatic stress component is zero) have no effect on the swelling. The total effect of the uniaxial compressive stress therefore arises from the hydrostatic pressure component, which produces a reduction in the swelling. Conversely, by the same argument, a tensile stress, for which the hydrostatic pressure component is negative, would be expected to lead to an increase in swelling.

These ideas may be generalized and applied to other systems, including those for which the Flory–Huggins theory is not appropriate. In this way it has been found possible, for example, to account for the observed increase in water content of cellulose and hair on the application of a tensile stress (Treloar 1953).

7.12. Swelling under torsional strain

The application of the general relations between swelling and strain developed in § 7.10 is not restricted to problems in which the state of strain is homogeneous throughout the specimen. An interesting example of a case involving inhomogeneous strain is provided by the torsion of a circular cylinder. In this case (Treloar 1972) the state of strain varies with radial position, and the degree of swelling also varies in a similar manner. On account of the complexity of the problem a general analytical solution is possible only for small torsional strains; for large strains numerical methods of computation have to be employed.

For a cylinder of radius a_0 in the unswollen state subjected to an axial extension in the ratio β_3 (referred to the unswollen axial length) and torsion ψ, in radians per unit *swollen* length, the analytical expression for the change ΔV in the swollen volume due to the torsion alone is

$$\frac{\Delta V}{V} = \frac{(\rho V_1/M_c)\beta_3\psi^2 a_0^2}{4\{2\chi - 1/(1-v_2)\}v_2^2}, \qquad (7.41)$$

where V is the total volume of the swollen cylinder. This expression is strictly valid only for an infinitesimal torsional strain, in which case the volume fraction of polymer (v_2) varies only to an infinitesimal extent across the section of the cylinder. It is to be noted that the predicted change of swelling $(\Delta V/V)$ is a *second-order* effect, being proportional to the *square* of the torsional strain, and that for a good swelling agent $(\chi < \tfrac{1}{2})$ it is negative, i.e. the swelling *decreases* on twisting. Fig. 7.13 shows a typical experimental result obtained by Loke, Dickinson, and Treloar (1972), in which the reduction of swelling $(-\Delta V/V)$ is plotted against the square of the torsion. This figure includes for comparison the small-strain solution represented by eqn (7.41), and the more accurate result obtained by numerical computation. On account of the smallness of the effect, the experimental accuracy

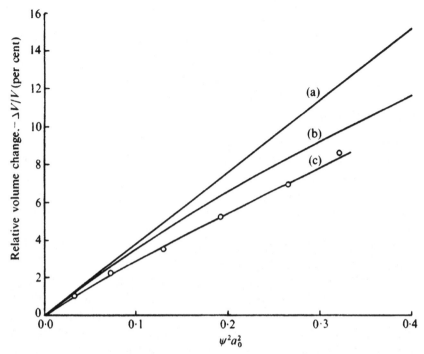

FIG. 7.13. Effect of torsion on swelling of cylinder. (a) Approximate theory (eqn (7.41)); (b) exact theory, numerical computation; (c) experimental.

was not good, but there is no doubt that the results substantiate the general form of the theoretical predictions. For a range of differently cross-linked rubbers, the quantitative deviations of the experimentally observed values of the relative volume change from the predictions of the accurate theory varied from 12 per cent to 23 per cent.

The reduction of swelling due to torsion has its origin in the presence of a normal component of compressive stress in the radial direction, discussed in detail in Chapter 12, where it is shown that this component of stress is proportional to the *square* of the torsional strain, unlike the shear stress, which is proportional to the first power of the strain. Being proportional to the square of the strain, the radial component of stress does not appear in the classical theory of elasticity, which of course is limited to infinitesimal strains. These observations on swelling in torsion are therefore of fundamental interest in demonstrating one of the important conclusions derivable only from the theory of large elastic deformations.

8

CROSS-LINKING AND MODULUS

8.1. Introduction

IN considering the extent to which the statistical theory is capable of representing the properties of a rubber in a quantitative manner attention has so far been concentrated on the *form* of the stress–strain relations for various types of strain. We have seen (Chapter 5) that to a first approximation the theoretical stress–strain relations, which involve only a single elastic constant G, provide an adequate basis for the interpretation of the properties of an actual rubber (provided that the strains do not approach the limiting extensibility of the network) though significant deviations are also observed, particularly in the case of simple extension. The further question which remains to be examined is that of the relation between the observed value of the shear modulus G and the value predicted by the statistical theory, which in its simplest form is given by

$$G = NkT = \rho RT/M_c, \tag{8.1}$$

where N is the number of network chains per unit volume and M_c is the corresponding 'chain molecular weight'.

The number of chains per unit volume is determined by the number of cross-linkages per unit volume introduced in the process of vulcanization. For an ideal network containing no network defects in the form of 'loose ends', intramolecular cross-linkages, etc., we have for the simplest form of cross-linkage (tetrafunctional), in which each junction point is the meeting point of four chains,

$$N = 2\nu, \tag{8.2}$$

where ν is the number of cross-links per unit volume, or the cross-link density. If, therefore, we can introduce a known number of cross-linkages it is a simple matter in principle to calculate the value of the modulus and to compare this with the directly measured value.

Unfortunately, the achievement of this objective is beset with difficulties on both the experimental and theoretical sides. On the experimental side the choice of a suitable chemical reaction to yield quantitative cross-linking has presented some difficulty, though this problem appears now to have been largely overcome. A more serious problem, which has not yet been fully resolved, arises from the deviations in the form of the force–extension curve from the theoretical form, as a result of which it is not possible to assign a unique and unambiguous value to the modulus.

On the theoretical side a completely satisfactory treatment of the effect of network imperfections is lacking. Apart from this there remain unresolved difficulties in the derivation of a precise theoretical formula for the modulus. Both these difficulties have already been discussed in Chapter 4. Practical studies have mostly been based on the formula (8.1) given by the elementary theory, together with the Flory 'loose-end' correction (eqn (8.3)) or some modification of it.

8.2. Early work

Sulphur-vulcanized rubbers

One of the first to examine the implications of the theory was Gee (1947), who compared the measured moduli for a series of natural rubber vulcanizates with the values to be expected on the basis of the amount of chemically combined sulphur, assuming one cross-link for each atom of combined sulphur. In general, the measured moduli were considerably lower than the calculated values, the differences being greater for compounds vulcanized with sulphur alone (where it amounted to a factor of about 5) than for those which incorporated an accelerator in addition to sulphur. These experiments showed clearly that no simple relation existed between the modulus and the amount of combined sulphur. This was attributed to the formation of polysulphide cross-linkages and other types of combination of the sulphur not involving inter-chain linkages—an explanation which was supported by the direct chemical evidence then existing (Bloomfield 1946; Farmer and Shipley 1946) and which has since been amply confirmed (Porter 1967).

Butyl rubbers

In an examination of butyl rubbers Flory (1946) took into account the effect of loose ends, as represented by his formula

$$G = (\rho RT/M_c)(1 - 2M_c/M). \tag{8.3}$$

The butyl rubbers are based on polyisobutylene:

$$[-C(CH_3)_2-CH_2-]_n,$$

in which is incorporated a small percentage of a conjugated diene (isoprene or butadiene) to provide the necessary double bonds for the cross-linking reaction with sulphur. A series of fractions of varying initial molecular weight M, each containing the same proportion of diolefin, were prepared and subsequently cross-linked under identical conditions; the number of cross-linkages per unit volume could therefore be assumed to be the same for all. The modulus measurements were made on the 'gel' fraction, obtained by dissolving out the 'sol' fraction consisting of molecules not chemically linked to the continuous network. The amount the sol fraction, for a given degree of cross-linking, increases as the initial molecular weight decreases, becoming 100 per cent at a critical value of molecular weight M' corresponding to the 'gel point', below which no true network formation occurs. On the theory of random cross-linking the number of cross-links at the gel point is equal to one-half the number of primary molecules.

Fig. 8.1 shows a typical result in the form of a plot of modulus $f/(\lambda - 1/\lambda^2)$ measured at $\lambda = 4 \cdot 0$ against the reciprocal of the measured molecular weight of the primary molecules. Assuming M_c

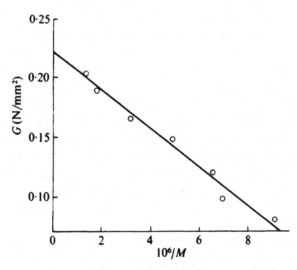

FIG. 8.1. Modulus G vs molecular weight M before vulcanization. (From Flory 1946.)

to be constant, this is consistent with the expected form of dependence of modulus on M (see eqn (8.3)). From the slope of the line a calculated value of M_c of 30 000 was obtained; this was consistent with the value calculated from the gel point ($M = M'$), namely, 35 000. However, the value of M_c calculated from the numerical value of the modulus extrapolated to $1/M = 0$ (for which the loose-end correction factor vanishes) was only 10 500, which is lower than the above by a factor of 3·5. The discrepancy fell slightly with increasing cross-linking, to a factor of 2·5 for an M_c (calculated from the modulus) of 8000.

Diazo compounds as cross-linking agents

The above results illustrate the difficulties encountered when indirect methods are used for the estimation of the degree of cross-linking. In later experiments Flory and co-workers employed a more direct method, based on the use of compounds of the type

$$R \begin{cases} O-CO-N{=}N-CO-O-CH_3 \\ O-CO-N{=}N-CO-O-CH_3 \end{cases}$$

known as bisazodicarboxylates, which react with the polymer at each of the $N{=}N$ bonds to give a calculable number of cross-linkages. In measurements on both natural and butadiene–styrene rubbers cross-linked by this means (Flory, Rabjohn, and Shaffer 1949) care was taken to achieve equilibrium under the applied stress; for this purpose a solvent absorption–desorption technique was used. The results for natural rubber are shown in Fig. 8.2, in which the (nominal) stress is plotted against the equivalent percentage of cross-linking agent, the values calculated from eqn (8.1) being represented by the dotted line. Though the departures from proportionality are significant, the experimental values are of the correct order of magnitude. Similar data were obtained for the butadiene–styrene rubbers.

This early work of Flory and his associates has been of great importance in establishing the consistency of the statistical theory in providing at least an approximately correct basis for the calculation of the numerical value of the elastic modulus of a cross-linked

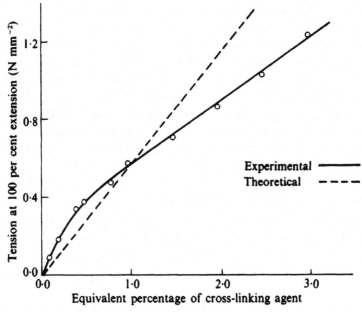

FIG. 8.2. Dependence of tensile force at 100 per cent extension on degree of cross-linking. (Flory *et al.* 1949.)

rubber. Measurements of the modulus, however, were based on the value of the force at an arbitrarily chosen extension, together with the assumption that the theoretical form of force–extension relation, represented by eqn (4.19a), is applicable. In practice, owing to the deviations from this relation noted above, the 'experimental' value of modulus will be dependent on the particular degree of extension employed in the measurements. Moreover, it cannot be assumed that the effect of these deviations will remain relatively unchanged as the degree of cross-linking is varied.

8.3. The experiments of Moore and Watson and of Mullins

In an attempt to overcome these difficulties a comprehensive series of experiments was undertaken by Moore and Watson, in conjunction with Mullins. Moore and Watson (1956) employed as cross-linking agents certain organic peroxides (e.g. di-(*t*-butyl) peroxide) which produce direct C—C cross-linkages by a catalytic reaction, thus avoiding the introduction of unwanted non-rubber constituents into the network structure. The reaction is quantitative, and hence the number of cross-linkages introduced can be determined by analysis of the reaction products. The material used was deproteinized natural rubber. Samples of different initial molecular weight before cross-linking were prepared by milling, and each of these samples was cross-linked to varying extents to produce a range of M_c values in the final state.

The deviations from the statistical theory were avoided by working with rubbers in the highly swollen state (or by extrapolation to this state), where the C_2 term in the Mooney equation becomes negligible (cf. Chapter 5). Measurements were made of the equilibrium swelling in n-decane; from these values the corresponding M_c-values were obtained by application of the Flory–Huggins relation (7.22). This relation contains the adjustable parameter χ, the value of which was obtained from the relation between modulus and swelling obtained by Mullins (1956) as represented in Fig. 7.7 (p. 146). In effect, therefore, the chemically estimated degree of cross-linking can be compared with the physical estimate of the same quantity derived from the modulus through eqn (8.1); for this particular purpose the use of the Flory–Huggins relation is merely an intermediate stage introduced as a convenient means of handling the experimental data.

The experiments of Mullins (1956) have already been dealt with in Chapter 7, but the correction which he introduced to take account of 'loose ends' has not been discussed. In considering this question we note first that Mullins obtained the values of C_1 for the swollen rubbers on the basis of the modified Mooney relation

$$f = 2v_2^{-\frac{1}{3}}(\lambda - 1/\lambda^2)(C_1 + C_2/\lambda), \qquad (8.4)$$

where f is the force per unit unstrained unswollen area. This equation implies that the constant $2C_1$ may be identified with the shear modulus G of the statistical theory; it therefore represents the modulus not of the actual swollen rubber (which for the case $C_2 = 0$ would be $2C_1 v_2^{-\frac{1}{3}}$) but of a hypothetical dry rubber for which the relation between modulus and swelling (v_2) is in accordance with the statistical theory. Hence, for an ideal network containing no loose ends we should have

$$2C_1 = G = \rho RT/M_c. \qquad (8.5)$$

Introduction of the Flory loose-end correction (eqn (8.3)) then gives

$$C_1 = \frac{\rho RT}{2M_c} - \frac{\rho RT}{M}. \qquad (8.6)$$

For a given value of M_c, i.e. for a series of rubbers of identical cross-linking, Mullins found a linear relation between C_1 and $1/M$ to apply; by extrapolation to $1/M = 0$ he thus obtained the value C_1^∞ corresponding to infinite initial molecular weight M. Putting $C_1^\infty = \rho RT/2M_c$, eqn (8.7) may be written in the form

$$C_1^\infty - C_1 = \rho RT/M, \tag{8.7}$$

which does not involve M_c explicitly; the data for the various series of rubbers of different M_c could thus be represented by a single line, as shown in Fig. 8.3. The slope of this plot, however, is not

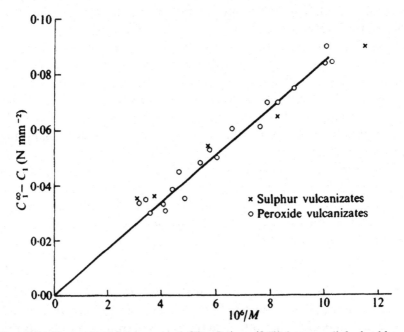

FIG. 8.3. Change in elastic constant $C_1^\infty - C_1$ (eqn (8.7)) for cross-linked rubbers, plotted against reciprocal of initial molecular weight M. (Mullins 1956.)

consistent with (8.7), but has the value $8 \cdot 5 \times 10^3$ N mm^{-2}, which is about 3·7 times the theoretical value ρRT. (This result might appear to be inconsistent with Flory's data for butyl rubbers, for which the fractional reduction in modulus was in good agreement with the theory; however, for a proper comparison, the *absolute* reduction in modulus should be calculated. This, as Mullins points out, would introduce a factor of about 3.)

It is interesting to note that the sulphur vulcanizates gave the same numerical dependence on $1/M$ as the peroxide vulcanizates.

The relation between modulus and degree of swelling shown in Fig. 7.7 (p. 146) was based on the values of C_1^{∞} obtained by the above empirical procedure. As observed in Chapter 7, these data were fitted to the Flory–Huggins equation (7.22) by choosing the value 0.41_3 for χ; this was the value used by Moore and Watson in the calculation of M_c from their equilibrium swelling data, a similar correction to that used by Mullins being introduced to take account

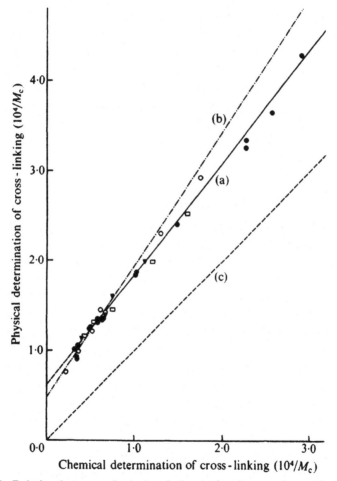

FIG. 8.4. Relation between physical and chemical estimates of cross-linking, as represented by $1/M_c$, for peroxide-vulcanized rubbers. (a) Experimental, using eqn (7.22); (b) experimental, using eqn (7.22a); (c) theoretical. The various symbols refer to different vulcanization conditions. (Moore and Watson 1956.)

of 'loose ends'. In this way values of the 'physical' estimate of cross-linking, represented by $1/M_c$, were obtained for comparison with the chemical estimates of the same quantity derived on the assumption that the number of chains is equal to twice the number of cross-links. The data so obtained are reproduced in Fig. 8.4, curve (a). Calculations were also made on the basis of Flory's modified formula (7.22a) (curve (b)). Taking the original formula to be the more appropriate, it is seen that the experimental data fall on a line whose slope is somewhat higher than that required theoretically. A more serious discrepancy, however, is the presence of a finite intercept of the experimental line on the vertical axis, which implies that the rubber behaves physically as if it possesses a finite number of cross-links, even in the absence of any specifically introduced chemical cross-links. Moore and Watson suggested that this effect could arise from the presence of entanglements between chains, such entanglements being equivalent to 'physical' cross-linkages. To account for the observations the number of such 'physical' cross-linkages would be relatively small, namely, 1 per 245 isoprene units.

8.4. Effect of entanglements

The above suggestion has been more closely examined in the later paper of Mullins (1959), already referred to in Chapter 7. In this paper the values of C_1 were derived from the *dry* rather than the swollen rubbers; they correspond, therefore, only to the C_1 contribution to the modulus. These values were some 10 per cent lower than the values previously derived from the swollen rubbers; in consequence, a slightly different value of χ (namely, 0·42) was required to fit the swelling data. The effect of these modifications on the final relation between physical and chemical estimates of cross-linking was, however, rather slight, as can be seen from Fig. 8.5.

In dealing with the question of physical entanglements Mullins argues that such entanglements will contribute to the network elasticity only if they are permanently 'entrapped' by adjacent chemical cross-links; the effective number of such entanglements will therefore increase with increasing cross-linkage. Furthermore, the chain-end correction should be based not on the number of chemical cross-links alone, but on the total number of chemical cross-links and physical entanglements. From this standpoint he

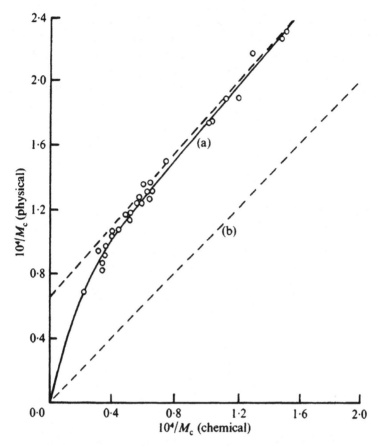

FIG. 8.5. Relation between physical and chemical estimates of cross-linking based on measurements of C_1^∞ for unswollen rubbers. (a) Experimental; (b) theoretical. (Mullins 1959a.)

writes, for the number of effective cross-links ν_e,

$$\nu_e = (\nu_{chem} + \nu_{ent})(1 - \beta M_c/M), \qquad (8.8)$$

in which ν_{chem} and ν_{ent} represent the respective numbers of chemical and physical cross-links and β is an empirical constant derivable from experiment. This leads to the expression

$$C_1 = (C_1^{\infty *} + \alpha)(1 - \beta M_c^*/M), \qquad (8.9)$$

in which M_c^* is the molecular weight of the chains in the chemically cross-linked network, calculated on the value of C_1, (namely, $C_1^{\infty *}$) corresponding to $1/M = 0$. Taking the experimental value of β, namely 2·3, the data reproduced in Fig. 8.5 could be represented by eqn (8.9) with $\alpha = 0.078$ N mm^{-2}. This is apparent from Fig. 8.6, in which the values of $1/M_c^{\infty*}$ calculated from eqn (8.9), with these values of the constants, are plotted against the corresponding chemically estimated values. The agreement between the two estimates is evidently very close.

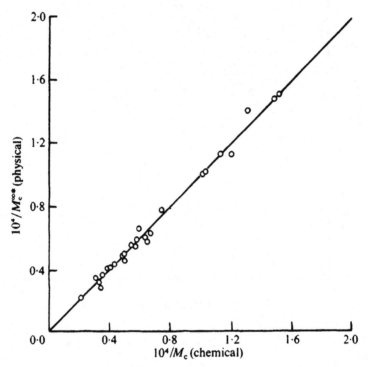

FIG. 8.6. Comparison of physical and chemical estimates of cross-linking after allowing for entanglements. (Mullins 1959a.)

It is interesting to note that Bobear (1966) applied the Mullins formulation with considerable success to data for radiation cross-linked silicone rubbers obtained by St. Pierre, Dewhurst, and Bueche (1959). A somewhat more elaborate treatment of the physical entanglement problem has also been developed by Meissner (1967).

8.5. Discussion and conclusion

While the experiments of Moore and Watson and of Mullins represent the most careful work on the problem of the relation between cross-linking and modulus yet carried out, there remain certain difficulties which are inherent in the problem and which it has not been possible to resolve in an entirely satisfactory manner. These difficulties stem from the deviations of the force–extension relation from the theoretical form referred to in the introduction to this chapter. The original treatment of Mullins (1956) is concerned with the modulus of the highly swollen rubber, for which the force–extension curve is of the theoretical form, i.e. $C_2 = 0$. In so far as the problem is regarded as relating the modulus, determined under the most appropriate conditions, to the chemically determined cross-link density, i.e. to the comparison of physical and chemical estimates of cross-linking, there can be no criticism of this procedure. Indeed, this was presumably the primary purpose of the investigation. It is in relating the modulus so determined to the properties and structure of the unswollen rubber that the difficulties arise. With regard to this problem there is no fundamental difference between the use of C_1-values determined in the swollen state and the derivation of the C_1-values from measurements on the unswollen rubber, as in the work of Mullins (1959b); the difference between the values obtained has been shown to be not large. By whichever method it is determined, the value of $2C_1$ may be very different from the small-strain shear modulus of the unswollen rubber, which is given by $2(C_1 + C_2)$. In Mullins' experiments the ratio C_2/C_1 varied from 0·43 to 0·79; the small-strain shear moduli would thus range from 1·43 to 1·79 times the values given by $2C_1$ alone. This is a very considerable difference, which of course was fully appreciated by the authors. Even if the arguments are limited to the C_1-value, however, the identification of the constant $2C_1$ with the NkT of the statistical theory is still open to question. This will be apparent from the examination of other types of strain, as given in Chapter 10, in which it is shown that the Mooney equation does not provide a satisfactory and self-consistent basis for the representation of the properties of a rubber under the most general type of strain. In particular, it is found that the empirical Mooney constant C_2 derived from simple extension data cannot be taken to represent the deviations from the form of strain–energy function given by the statistical theory.

This difficulty is not entirely disposed of by the observation that the C_1-values for the dry rubber differ only slightly from the corresponding values derived from measurements on the swollen rubber, for which the statistical theory is apparently valid. The hiatus in this argument is that the values of C_1 used by Mullins are not the actual values for the swollen rubber, but the values converted by the factor $v_2^{-\frac{1}{3}}$ (eqn (8.4)) so as to relate to the unswollen state. The values so obtained represent the properties not of the actual rubber but of a hypothetical rubber for which the relation between C_1 in the dry and in the swollen states is in accordance with the predictions of the statistical theory. There is no basis for this assumption other than the *experimental* observation that the values of C_1 for the unswollen rubber calculated in this way are consistent with the values of C_1 derived from measurements on the dry rubber. But if the latter are open to question (as a representation of network properties) the basis for this assumption is removed.

To sum up, therefore, we may say that in view of the fact that the properties of a rubber in the unswollen state cannot be described in terms of a single elastic constant it is not possible to obtain an unambiguous relation between the degree of cross-linking and *the* modulus, since no single value of modulus is applicable in the region of large strains. If attention is restricted to the values of C_1 derived either from the swollen or from the dry rubber, then the increase in modulus with increase in degree of cross-linking is within about 25 per cent of the value derived from the elementary statistical theory, but the absolute values of modulus differ rather more, owing to the finite intercept on the vertical axis corresponding to zero cross-linking. This intercept is probably connected with the presence of physical entanglements between chains, which act in a manner similar to chemical cross-linkages.

In attempting to assess the significance of these conclusions it is necessary to bear in mind the very general nature of the theoretical model with which the experimental results are being compared. This model is obviously a gross oversimplification of the physically existing network structure, and the elementary treatment, as we have seen in Chapter 4, introduces approximations into the treatment which are not capable of absolute justification. This is exemplified by the more elaborate treatment of James and Guth, which as noted in Chapter 4 suggests that the value of modulus given by the elementary theory may require modification by a factor of $\frac{1}{2}$. Whether or not this figure ultimately proves to be justified, the result may give some indication of the limitations on the quantitative accuracy of the theory in its present form. The experimental data, however interpreted, would thus appear to fall within the range of uncertainty of the theory itself, and to this extent may be regarded as in satisfactory agreement with the theory.

If this appears to be a somewhat negative conclusion, some satisfaction may be drawn from the fact that the prediction of the *absolute* magnitude of a physical property from considerations of a quite general character not involving the introduction of any parameters derived from experimental observations (other than Boltzmann's constant, obtained from observations on gases) is an achievement of considerable significance. Taking a broad view, the presence of relatively minor inconsistencies is of less importance than the major success of the statistical theory in predicting not only the approximate form of the stress–strain relations of a rubber in any state of strain, but, in addition, the absolute value of the modulus to within a small numerical factor.

PHOTOELASTIC PROPERTIES
OF RUBBERS

9.1. Refractive index and polarizability

IT is well known that many crystals are optically anisotropic, being characterized by different values of refractive index for different directions of propagation of light through them, or, more precisely, for different directions of polarization of the transmitted light. This optical anisotropy, which gives rise to the phenomenon of double refraction, has its origin in the different polarizabilities of the medium for different directions of the electric field or electric vector in the electromagnetic wave. Generally there is a close connection between the optical anisotropy and the mechanical or elastic anisotropy, since both are directly related to the type of symmetry exhibited by the molecular structure of the crystal. Materials such as glasses and rubbers, whose structure is essentially irregular or amorphous, are normally isotropic in their physical properties; such materials do not show double refraction in the unstrained state. But if such materials are deformed by the application of a stress their structural randomness is disturbed; they cease to be isotropic and begin to exhibit properties in some ways akin to those of a crystal. In particular they show double refraction or birefringence, which is a rather sensitive indicator of the structural dissymmetry induced by the strain.

The effect of an electric field on a polarizable medium is to produce a separation of charge or polarization, the *polarizability* being defined as the ratio of induced dipole moment to field strength. In an optically anisotropic medium the properties are represented by the refractive index ellipsoid, the principal axes of which represent the three principal refractive indices n_1, n_2, and n_3 in three mutually perpendicular directions. This in turn is related to the polarizability ellipsoid, whose axes correspond to the principal polarizabilities β_1, β_2, and β_3, and coincide in direction with the axes of the refractive index ellipsoid. In an optically isotropic medium the relation between the polarizability per unit volume (β) and the refractive index n is expressed by the Lorentz–Lorenz formula

$$\frac{n^2-1}{n^2+2} = \frac{4\pi}{3}\beta. \tag{9.1}$$

9.2. Optical properties of long-chain molecules

The problem of dealing with the optical properties of strained rubbers or *strain birefringence*, which was first successfully attacked by Kuhn and Grün (1942), requires the calculation of the principal polarizabilities in a strained cross-linked network. The method used by Kuhn and Grün follows closely on the lines taken in the treatment of the corresponding problem of the elastic or mechanical properties of the network, as discussed in Chapter 4. First, the optical properties of the single chain were determined as a function of the vector distance between its ends. Next, the contribution of an individual chain in the network to the total polarizability of the network, in any specified state of strain, was determined. Finally, the total polarizability of the network was obtained by summation over all the chains.

As in the treatment of the elastic properties, the actual molecule is replaced by a hypothetical chain of randomly jointed links of equal length. The optical properties are introduced by associating with each link an optical anisotropy, defined by polarizabilities α_1 in the direction of its length and α_2 in the transverse direction, the link being therefore assumed to possess axial symmetry.

Let us assume the chain, containing n links each of length l, to be held with its ends separated by a vector distance r, which for convenience may be taken to coincide with the axis $0x$ of a rectangular coordinate system. The resultant components of polarizability for the whole chain along the axes $0x$, $0y$, and $0z$ may be calculated if the directions of all the links are known; these directions will of course be determined by the specific conformation of the chain. However, if the number of links is large it is possible to define statistically the *distribution* of link angles and hence to obtain the corresponding *mean* components of polarizability for the chain.

The angular distribution of link directions has already been discussed in Chapter 6; it is represented by the expression

$$\mathrm{d}n = \frac{n\beta}{\sinh\beta} e^{\beta\cos\theta} \cdot \tfrac{1}{2}\sin\theta\,\mathrm{d}\theta \cdot \frac{\mathrm{d}\phi}{2\pi}, \tag{9.2}$$

in which β is the inverse Langevin function $\mathscr{L}^{-1}(r/nl)$. For a link defined by the angles θ (with respect to $0x$) and ϕ (with respect to the plane $x0y$), the components of the polarizability tensor referred to the coordinate axes may be shown to be

$$\alpha_{xx} = \alpha_1 \cos^2 \theta + \alpha_2 \sin^2 \theta,$$

$$\alpha_{yy} = (\alpha_1 - \alpha_2) \sin^2 \theta \cos^2 \phi + \alpha_2,$$

$$\alpha_{zz} = (\alpha_1 - \alpha_2) \sin^2 \theta \sin^2 \phi + \alpha_2,$$

$$\alpha_{xy} = \alpha_{yx} = (\alpha_1 - \alpha_2) \sin \theta \cos \theta \cos \phi, \qquad (9.3)$$

$$\alpha_{yz} = \alpha_{zz} = (\alpha_1 - \alpha_2) \sin^2 \theta \sin \phi \cos \phi,$$

$$\alpha_{zx} = \alpha_{xz} = (\alpha_1 - \alpha_2) \sin \theta \cos \phi \sin \phi,$$

the notation being such that α_{xy} is the polarizability in the direction $0x$ for a field applied in the direction $0y$.

If γ_{xx}, γ_{yy}, etc., are the corresponding components of polarizability for the whole chain, we then have six equations of the type

$$\gamma_{xx} = \int \alpha_{xx} \, dn, \qquad (9.4)$$

in which dn is the number of links in the angular range $d\theta, d\phi$. Substitution of the expression (9.2) for dn and integration with respect to the angular variables yields the result

$$\gamma_{xx} = n \left\{ \alpha_1 - (\alpha_1 - \alpha_2) \frac{2r/nl}{\mathscr{L}^{-1}(r/nl)} \right\}, \qquad (9.5a)$$

$$\gamma_{yy} = \gamma_{zz} = n \left\{ \alpha_2 + (\alpha_1 - \alpha_2) \frac{r/nl}{\mathscr{L}^{-1}(r/nl)} \right\}, \qquad (9.5b)$$

$$\gamma_{xy} = \gamma_{yz} = \gamma_{zx} = 0. \qquad (9.5c)$$

The fact that the cross-polarizabilities γ_{xy}, etc. in (9.5c) are all zero implies that the principal axes of the polarizability ellipsoid coincide with the axes of coordinates, while eqn (9.5b) implies that the two transverse polarizabilities are equal. The whole chain therefore has the optical properties of a uniaxial body whose optic axis coincides with the r vector, i.e. the line joining the ends of the chain. These properties would, of course, be anticipated from the basic symmetry of the system.

Writing γ_1 and γ_2 in place of γ_{xx} and γ_{yy} for the respective longitudinal and transverse polarizabilities of the whole chain, the optical anisotropy or difference of principal polarizabilities, from

(9.5a) and (9.5b), may be written

$$\gamma_1 - \gamma_2 = n(\alpha_1 - \alpha_2)\left\{1 - \frac{3r/nl}{\mathscr{L}^{-1}(r/nl)}\right\}. \tag{9.6}$$

It will be noted that the dependence of optical anisotropy on chain extension involves only the *fractional* chain extension r/nl. This means that the form of this dependence is the same for all chains. At the maximum chain extension $(r = nl)$ the anisotropy becomes $n(\alpha_1 - \alpha_2)$, corresponding to n links in full alignment.

The form of the function (9.6) is shown in Fig. 9.1 in terms of the *relative* chain anisotropy $(\gamma_1 - \gamma_2)/n(\alpha_1 - \alpha_2)$.

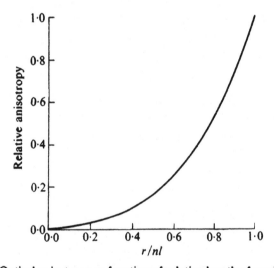

FIG. 9.1. Optical anisotropy as function of relative length of random chain.

By expansion of the inverse Langevin function in eqn (9.6) as in eqn (6.10a) (p. 107) we obtain the corresponding series expression

$$\gamma_1 - \gamma_2 = n(\alpha_1 - \alpha_2)\left\{\frac{3}{5}\left(\frac{r}{nl}\right)^2 + \frac{36}{175}\left(\frac{r}{nl}\right)^4 + \frac{108}{175}\left(\frac{r}{nl}\right)^6 + \dots\right\}. \tag{9.6a}$$

For values of (r/nl) which are not too large the first-term approximation

$$\gamma_1 - \gamma_2 = n(\alpha_1 - \alpha_2)\frac{3}{5}\left(\frac{r}{nl}\right)^2 \tag{9.6b}$$

is valid. This approximation, which is entirely analogous to the Gaussian approximation for the chain entropy (eqn (6.7b) p. 104), implies that for small chain extensions the optical anisotropy is proportional to the square of the chain vector length r. It is interesting to note, further, that for a free chain, for which $\overline{r^2} = nl^2$, the mean anisotropy is simply

$$\gamma_1 - \gamma_2 = \tfrac{3}{5}(\alpha_1 - \alpha_2), \tag{9.6c}$$

i.e. three-fifths of the anisotropy of a single statistical link.

A further observation of interest in practical applications is that the expression (9.6), unlike the analogous expression for the entropy, does not tend to infinity at high chain extensions; it is therefore more readily approximated by a finite number of terms in the expansion (9.6a). A simple alternative series, arrived at empirically, namely,

$$1 - \frac{3r/nl}{\mathscr{L}^{-1}(r/nl)} \simeq \frac{3}{5}\left(\frac{r}{nl}\right)^2 + \frac{1}{5}\left(\frac{r}{nl}\right)^4 + \frac{1}{5}\left(\frac{r}{nl}\right)^6 \tag{9.6d}$$

has been found to be numerically accurate to within 1 per cent over the whole range of r/nl from 0 to 1 (Treloar 1954), and is thus of considerable value in numerical calculations.

9.3. The Gaussian network

Simple extension

In their treatment of the optical properties of the Gaussian network Kuhn and Grün proceed by the following stages.

1. The network is assumed to contain N chains per unit volume whose r vectors in the unstrained state are distributed randomly in direction.
2. On deformation the components of vector length for each chain are assumed to change in the same ratio as the corresponding dimensions of the bulk rubber (affine deformation). This determines the length r' and angular coordinates (θ', ϕ') of the chain vector length in terms of the original length r and orientation (θ, ϕ) and the extension ratio λ.
3. The principal polarizabilities for a chain of length r' being given by eqns (9.5a) and (9.5b), the contribution of a given chain to the total polarizabilities referred to the principal axes of strain are obtained by means of equations of the type (9.3).

4. Integration over all chains then gives the total polarizabilities for the whole network.

Denoting the network polarizabilities respectively parallel and perpendicular to the direction of the extension by β_1 and β_2, the final result obtained after following through the above procedure is represented, in the first-term approximation corresponding to (9.6b), by the equations,

$$\beta_1 = N\left\{\frac{n}{3}(\alpha_1 + 2\alpha_2) + \tfrac{2}{15}(\alpha_1 - \alpha_2)\frac{\overline{r^2}}{nl^2}\left(\lambda^2 - \frac{1}{\lambda}\right)\right\},$$

$$\beta_2 = N\left\{\frac{n}{3}(\alpha_1 + 2\alpha_2) - \tfrac{1}{15}(\alpha_1 - \alpha_2)\frac{\overline{r^2}}{nl^2}\left(\lambda^2 - \frac{1}{\lambda}\right)\right\},$$

(9.7)

where $\overline{r^2}$ is the mean-square vector length in the unstrained state.

The final step involves transforming these polarizabilities into the corresponding principal refractive indices n_1 and n_2. For this purpose Kuhn and Grün assume the validity of the Lorentz–Lorenz relation (9.1) in respect of the separate polarizabilities—an assumption which is not strictly justified, but which is probably sufficiently accurate, in view of the relative smallness of the *differences* of polarizabilities compared with the mean value. Introducing the approximation

$$\frac{n_1^2 - 1}{n_1^2 + 2} - \frac{n_2^2 - 1}{n_2^2 + 2} \simeq \frac{6n_0}{(n_0^2 + 2)^2}(n_1 - n_2),$$

(9.8)

in which n_0 is the mean refractive index, given by $\tfrac{1}{3}(n_1 + 2n_2)$, and putting $\overline{r^2} = nl^2$, as for a corresponding set of free chains, they finally obtain the optical anisotropy in the form

$$n_1 - n_2 = \frac{2\pi}{45}\frac{(n_0^2 + 2)^2}{n_0}N(\alpha_1 - \alpha_2)\left(\lambda^2 - \frac{1}{\lambda}\right).$$

(9.9)

The quantity $n_1 - n_2$, representing the strain birefringence, is thus determined by the number of chains per unit volume (N), the anisotropy of the random link in the chain $(\alpha_1 - \alpha_2)$, and the extension ratio λ. It is not directly affected by the number of links n in the chain.

Stress-optical coefficient

To obtain the relation between the birefringence and the applied *stress* it is necessary to introduce the Gaussian stress–strain relation as already derived from the network theory. (At this point it is necessary to go beyond the original presentation of Kuhn and Grün, which preceded this development of the network theory.) For the (true) stress t we have (p. 67)

$$t = NkT(\lambda^2 - 1/\lambda). \tag{9.10}$$

It is seen that both the birefringence and the stress involve identical functions $(\lambda^2 - 1/\lambda)$ of the strain. We may therefore eliminate this variable and obtain the linear stress–birefringence relationship

$$n_1 - n_2 = Ct, \tag{9.11}$$

in which the strain-independent parameter C is given by

$$C = \frac{2\pi}{45kT} \frac{(n_0^2 + 2)^2}{n_0} (\alpha_1 - \alpha_2). \tag{9.12}$$

Eqn (9.11), which states that the birefringence is directly proportional to the stress, is equivalent to Brewster's law, previously found to apply to glassy materials under small-strain conditions. The constant C is known as the *stress-optical coefficient*. It is important to note that this constant depends only on the mean refractive index (which is not in itself a network property) and on the optical anisotropy of the random link. It does not involve N, and is therefore independent of the degree of cross-linking of the network.

General homogeneous strain

Kuhn and Grün's analysis was limited to the case of simple extension, in which the optical properties correspond to those of a uniaxial crystal whose optic axis coincides with the direction of extension, and are specified by two refractive indices n_1 and n_2. In the more general problem of the pure homogeneous strain the refractive index ellipsoid will have three unequal axes n_1, n_2, and n_3 corresponding to the principal axes λ_1, λ_2, and λ_3 of the strain ellipsoid.

This problem has been treated by the author (Treloar 1947), who used an essentially similar but mathematically somewhat simplified model compared with that used by Kuhn and Grün for the purpose. The network of N chains was divided up into $N/3$ sets of three mutually perpendicular chains, each having the same value of r in the unstrained state. For any one set the principal polarizabilities were calculated under an imposed strain corresponding to principal extensions λ_1, λ_2, and λ_3. The expressions for the resultant polarizabilities were shown to be independent of the orientation of the axes of the strain ellipsoid with respect to the original r vectors, and hence applied equally to all the sub-sets. The principal polarizabilities β_{xx}, β_{yy}, β_{zz}, for the whole network were thus readily obtained. These are given by

$$\beta_{xx} = N\{\tfrac{1}{3}n(\alpha_1+2\alpha_2)+\tfrac{1}{15}(\alpha_1-\alpha_2)(2\lambda_1^2-\lambda_2^2-\lambda_3^2)\},$$

$$\beta_{yy} = N\{\tfrac{1}{3}n(\alpha_1+2\alpha_2)+\tfrac{1}{15}(\alpha_1-\alpha_2)(2\lambda_2^2-\lambda_3^2-\lambda_1^2)\}, \qquad (9.13)$$

$$\beta_{zz} = N\{\tfrac{1}{3}n(\alpha_1+2\alpha_2)+\tfrac{1}{15}(\alpha_1-\alpha_2)(2\lambda_3^2-\lambda_1^2-\lambda_2^2)\}.$$

Converting to refractive indices through eqn (9.1) and introducing the approximation (9.8), the *differences* of refractive index are obtained in the form

$$n_1 - n_2 = \frac{(n_0^2+2)^2}{n_0}\frac{2\pi N}{45}(\alpha_1-\alpha_2)(\lambda_1^2-\lambda_2^2), \qquad (9.14)$$

with corresponding expressions for $n_2 - n_3$ and $n_3 - n_1$. The difference $n_1 - n_2$ represents the birefringence for a ray of light propagated in the λ_3-direction, the n_1 and n_2 corresponding to the two principal directions of electric vector (planes of polarization) for such a ray. From eqn (9.14) it is seen that the birefringence is proportional to the difference of the *squares* of the corresponding principal extension ratios in a plane perpendicular to the ray direction.

The principal stress–strain relations derived from the Gaussian network theory (p. 67) are of the form

$$t_1 - t_2 = NkT(\lambda_1^2 - \lambda_2^2). \qquad (9.15)$$

We have, therefore,

$$n_1 - n_2 = C(t_1 - t_2), \qquad (9.16)$$

in which, as before, the constant C is the stress–optical coefficient defined by (9.12). It follows that the birefringence for light propagated along one of the principal axes of strain is proportional to the *difference of principal stresses in the transverse plane*. This is a generalization of Brewster's law as previously applied to the case of uniaxial strain. The result (9.15) has no analogue in the classical

theory of photoelasticity but reduces to the classical form (birefringence proportional to difference of principal strains) when the
deformations are small.

9.4. The effect of swelling

It is a simple matter to extend the theory to the case of a rubber
swollen by a solvent. For this purpose it is assumed that the solvent
is optically neutral and isotropic, even when the rubber is strained.
The solvent is therefore assumed to affect the optical properties
only indirectly, by altering the mean chain extensions. In the
unstrained state the r.m.s. chain extension $(\overline{r^2})^{\frac{1}{2}}$ is assumed to be
proportional to the linear dimensions of the swollen rubber, i.e., to
$v_2^{-\frac{1}{3}}$, where v_2 is the volume fraction of rubber in the mixture. For a
network of chains occupying unit volume in the unswollen state the
effect on the polarizabilities is then simply to alter the mean value of
$\overline{r^2}$ in the expressions for the polarizabilities (corresponding to eqns
(9.7) for the case of simple extension) from nl^2 to $nl^2 v_2^{-\frac{2}{3}}$. A further
factor of v_2 has to be introduced to relate the polarizabilities to unit
volume in the swollen state. The modified form of (9.14) is therefore

$$n_1 - n_2 = \frac{(n_0^2 + 2)^2}{n_0} \cdot \frac{2\pi N}{45} (\alpha_1 - \alpha_2) v_2^{\frac{1}{3}} (\lambda_1^2 - \lambda_2^2), \qquad (9.17)$$

where n_0 is now the mean refractive index of the swollen rubber,
which effectively includes the contribution of the isotropic polarizability of the solvent. In so far as the refractive index of the solvent
differs from the refractive index of the (unstrained) rubber, the
factor $(n_0^2 + 2)^2 / n_0$ in eqn (9.17) will have some effect on the
birefringence.

The effect of swelling on the principal stresses has been given in
Chapter 4. The differences of principal stresses are represented by
equations of the type

$$t_1 - t_2 = NkT v_2^{\frac{1}{3}} (\lambda_1^2 - \lambda_2^2). \qquad (9.18)$$

Since the factor $v_2^{\frac{1}{3}}$ occurs in both (9.17) and (9.18), the stress-optical
coefficient C, as represented by eqn (9.12), remains unchanged,
except for the possible effect of the solvent on n_0, noted above.

9.5. The non-Gaussian network

The foregoing results are applicable only to the *Gaussian* network and are therefore valid only in so far as the chains are

sufficiently long and the strains are not too large. The treatment of the optical properties of the non-Gaussian network, like that of the mechanical properties, is considerably more complex. Kuhn and Grün (1942) attacked this problem by an extension of their original treatment, discussed in § 9.3 above. The assumption of an 'affine' displacement of network junction points was retained, but in place of the first-term approximation (9.6b) for the optical anisotropy of the single chain the first three terms in the series (9.6a) were included. The resulting expressions for the network polarizabilities (comparable to 9.7), in the Gaussian approximation) thus contain additional terms involving $\overline{r^4}/n^2l^4$ and $\overline{r^6}/n^3l^6$. To evaluate these terms Kuhn and Grün assumed the distribution of r vectors in the unstrained state to be Gaussian, as for a corresponding set of free chains, and therefore wrote

$$\overline{r^2}/nl^2 = 1; \qquad \overline{r^4}/n^2l^4 = 5/3; \qquad \overline{r^6}/n^3l^6 = 35/9. \qquad (9.19)$$

With these values the expression for the anisotropy of network polarizability for a simple extension in the ratio λ becomes

$$\beta_1 - \beta_2 = N(\alpha_1 - \alpha_2)\left\{\frac{1}{5}\left(\lambda^2 - \frac{1}{\lambda}\right) + \frac{2}{175n}\left(6\lambda^4 + 2\lambda - \frac{8}{\lambda^2}\right) + \right.$$
$$\left. + \frac{6}{875n^2}\left(10\lambda^6 + 6\lambda^3 - \frac{16}{\lambda^3}\right) + \ldots\right\}. \qquad (9.20)$$

As in the case of the mechanical properties, the higher-order terms in this expression involve n, the number of links in the randomly jointed chain. However, as we have already seen, the assumption of a distribution of r vectors corresponding to an assembly of free chains is open to criticism; it is also incompatible, in the non-Gaussian treatment, with the assumption of an affine displacement of junction points (Treloar 1954). It may well be nearer the truth to assume that all the chains have the same initial vector length. If this is assumed to be given by $r^2 = nl^2$ we should then have

$$\overline{r^4}/n^2l^4 = \overline{r^6}/n^3l^6 = 1. \qquad (9.19a)$$

Inserting these values into Kuhn and Grün's equations in place of the values (9.19), and introducing the closer approximation (9.6d) in place of the series (9.6a) for the chain anisotropy, the author

(Treloar 1954) obtained the result

$$\beta_1 - \beta_2 = N(\alpha_1 - \alpha_2)\left\{\frac{1}{5}\left(\lambda^2 - \frac{1}{\lambda}\right) + \frac{1}{150n}\left(6\lambda^4 + 2\lambda - \frac{8}{\lambda^2}\right) + \right.$$
$$\left. + \frac{1}{350n^2}\left(10\lambda^6 + 6\lambda^3 - \frac{16}{\lambda^3}\right)\right\}, \quad (9.21)$$

which differs from (9.20) in assigning smaller numerical values to the coefficients in the higher-order terms. The difference is illustrated in Fig. 9.2 (curves (a) and (b)) for the case $n = 25$. Both expressions reduce to the Gaussian form if the higher-order terms are neglected (cf. eqn (9.7)).

An alternative mathematical treatment of the stress-optical properties of the non-Gaussian network, based on a somewhat similar physical model, has been developed by Smith and Puett (1966).

The effect of relaxing the affine deformation assumption has been examined by the author (Treloar 1954) using as a basis the Flory–Rehner tetrahedral cell model (p. 117). The results (which cannot be represented by general mathematical formulae) are shown in Fig. 9.2, in terms of the 'optical orientation factor', or anisotropy relative to that of the fully extended chains. It is seen that the conclusions are not greatly affected according to whether the central junction point was subjected to an affine displacement or allowed to assume its equilibrium position under the action of the forces in the adjacent chains. As noted previously, however, the restriction to an affine displacement had the effect of reducing the extensibility.

The relation between the optical anisotropy and the stress, for two different values of n, based on the tetrahedral cell model with affine displacement, is shown in Fig. 9.3. Whereas in the Gaussian approximation this relation is linear (Brewster's law) and independent of n, the non-Gaussian theory leads to departures from this behaviour which become apparent at smaller values of the stress the smaller is the value of n (i.e. of chain length). For sufficiently small values of n even the slope at the origin (or limiting value of the stress-optical coefficient) is dependent on n.

A more precise theory of the photoelastic properties of the non-Gaussian network would need to find some method of dealing with the problem of the distribution of chain vector lengths in the

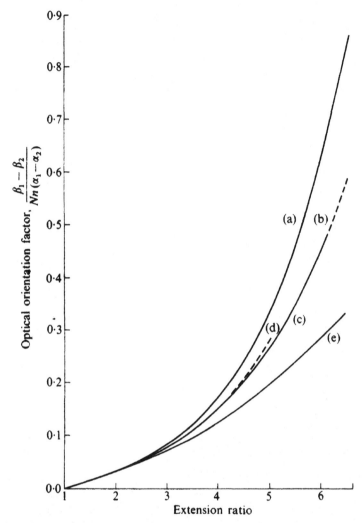

FIG. 9.2. Theoretical optical anisotropy for network, with $n = 25$. (a) Kuhn–Grün theory, eqn (9.20); (b) Kuhn–Grün theory, modified, eqn (9.21); (c) four-chain model, non-affine deformation; (d) four-chain model, affine deformation; (e) Gaussian.

unstrained state, which would necessarily include also the consideration of the distribution of chain contour lengths. In the absence of such a theory the conclusions derived from the rather crude models referred to above cannot be regarded as quantitatively precise, though they do indicate the general form of the dependence of optical anisotropy on strain and on the parameter n, the number of statistical links in the chain.

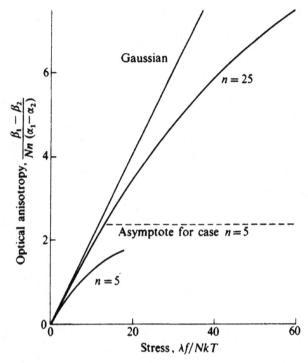

FIG. 9.3. Optical anisotropy as function of stress, derived from four-chain model, with $n = 5$ and $n = 25$.

9.6. Measurement of birefringence

The theoretical relations derived in the foregoing sections are concerned with the values of the three refractive indices n_1, n_2, and n_3 corresponding to the principal axes of the refractive index ellipsoid, whose axes correspond in direction with the axes of the strain ellipsoid. If these axes are chosen so as to coincide respectively with the axes Ox, Oy, and Oz of a rectangular coordinate system, then the refractive index of the medium for a ray of light travelling in the direction Ox will be n_2 if it is polarized with its electric vector parallel to Oy and n_3 if its electric vector is parallel to Oz. For any other plane of polarization the ray may be resolved into two components, having refractive indices n_2 and n_3. The value of $n_2 - n_3$ is a measure of the birefringence for the direction of propagation Ox. The corresponding birefringences for directions of propagation Oy and Oz are respectively $n_3 - n_1$ and $n_1 - n_2$.

In the case of simple extension the system possesses uniaxial symmetry, and is characterized by one optic axis. For light propagated along this axis the birefringence is zero. For the general (triaxial) strain the properties correspond to those of what is historically defined as a 'biaxial' crystal, this term referring to the fact that in such a system there are two (and only two) directions of propagation, called the optic axes, along which there is no birefringence. These directions are those for which the section of the refractive index ellipsoid normal to the ray direction are circular (Jenkins and White 1957).

The measurement of birefringence requires the use of an optical compensator, which introduces an adjustable optical path difference between the two rays polarized in perpendicular planes to compensate for their different refractive indices in the medium (Fig. 9.4). The experimental arrangement consists of a polarizer P (Nicol

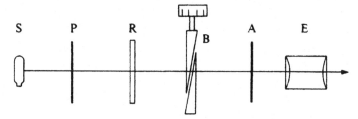

FIG. 9.4. Experimental arrangement for measurement of double refraction. S: light-source; P: polarizer; R: specimen; B: Babinet compensator; A: analyser; E: eyepiece.

prism or polaroid plate) set so that the light incident on the specimen R is polarized at an angle of 45° to the directions of the principal axes of strain, a compensator B, and an analysing Nicol or polaroid set at right angles to the polarizer. The principle of the measurement is indicated in Fig. 9.5. The incident light, polarized in the direction OP, is resolved into two components, Oa and Ob, parallel to the principal axes of strain in the specimen. These travel with velocities c/n_1 and c/n_2, respectively, where c is the velocity of light *in vacuo*. After passing through the analyser A, these have components Oa' and Ob' of equal amplitudes, polarized in the same plane. For an optically isotropic material these two components, being in opposite senses, would interfere to produce zero resultant intensity of illumination. For an optically anisotropic material, however, there will be an optical path difference Δ between the two

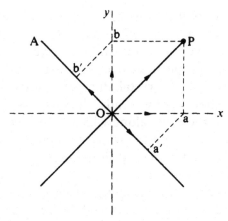

FIG. 9.5. Interference effects. OP: plane of polarizer; Ox, Oy: principal axes of strain; OA: plane of analyser.

components, which will in general lead to the transmission of light. Extinction will occur only when the path difference corresponds to an integral number of wavelengths. If d is the thickness of the specimen and λ_1 and λ_2 are the wavelengths in the medium corresponding to the refractive indices n_1 and n_2, then

$$\lambda_1 = \lambda_0/n_1 \quad \text{and} \quad \lambda_2 = \lambda_0/n_2, \tag{9.22}$$

where λ_0 is the vacuum wavelength. The path difference, in number of waves, is therefore given by

$$\Delta = \frac{d}{\lambda_1} - \frac{d}{\lambda_2} = (n_1 - n_2)\frac{d}{\lambda_0}. \tag{9.23}$$

The condition for extinction of the transmitted light is therefore,

$$(n_1 - n_2)d/\lambda_0 = k \qquad (k = 0, 1, 2, \ldots). \tag{9.23a}$$

One of the simplest forms of compensator is the Babinet compensator, which consists of two quartz wedges of equal wedge-angle, cut with their optic axes at right angles to each other, so as to introduce path differences in opposite senses (Fig. 9.4). At the point at which the two thicknesses are equal there appears a dark band, corresponding to $k = 0$, while on either side there are parallel bands corresponding to $k = +1, +2, \ldots$ and $k = -1, -2, \ldots$, etc., when the system is viewed in monochromatic light. Introduction of a birefringent specimen displaces the band system, which may then be brought back into its original position (with reference to a cross-wire) by displacement of the first wedge in the compensator through a measurable distance. Knowing the displacement corresponding to one wavelength, this immediately gives the equivalent optical path difference.

In practice it is convenient to work with white light so as to be able to identify the zero-displacement fringe, which then appears as a black fringe surrounded by a succession of coloured fringes, corresponding to the extinction of the successive constituent colours of the white light.

9.7. Investigations on natural rubber

A typical example of the behaviour of vulcanized natural rubber in simple elongation is shown in Fig. 9.6 (Treloar 1947). Both the

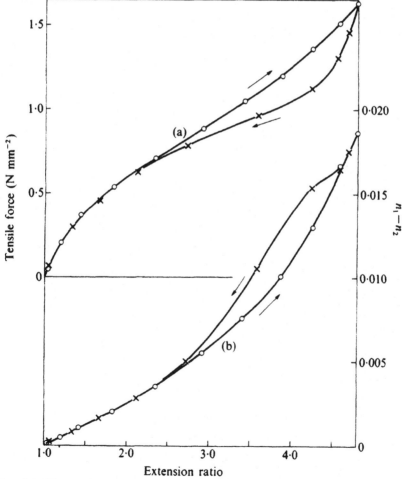

FIG. 9.6. Hysteresis in tension (a) and birefringence (b) curves for natural rubber.

tension and the birefringence curves show marked hysteresis loops in the same region of extension, but whereas the tension is lower in the retraction half-cycle, the birefringence is *higher*. Consequently, a plot of birefringence against stress yields a still more pronounced hysteresis loop (Fig. 9.7, 25 °C curve). However, at the lower stresses (corresponding to extension ratios below about 3·0) the expected proportionality between birefringence and stress (eqn (9.11)) is exactly realized.

It is inferred that these hysteresis loops are the result of crystallization, the crystallites themselves being oriented in the direction of

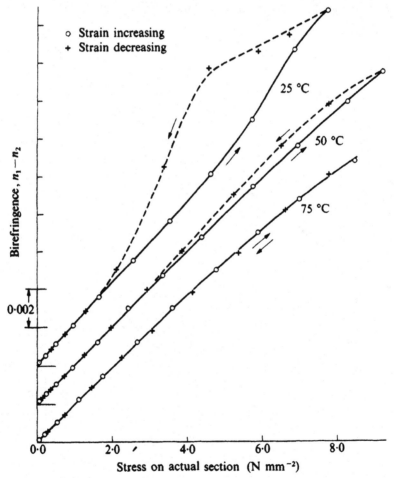

FIG. 9.7. Relation between birefringence and stress for natural rubber at 25 °C, 50 °C and 100 °C.

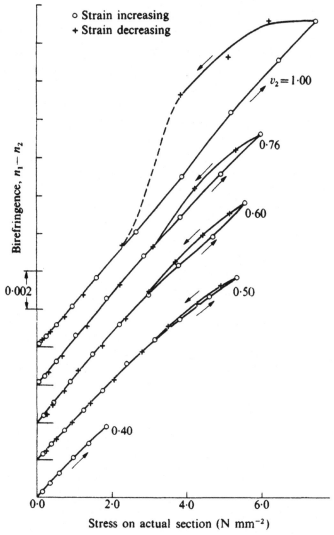

FIG. 9.8. Effect of swelling in toluene on hysteresis. v_2 = volume fraction of rubber.

the extension and therefore making a specific contribution to the birefringence which is additional to that due to the amorphous network. This supposition is confirmed by experiments at higher temperatures (Fig. 9.7) which show a progressive diminution and ultimate suppression of the effects with rising temperature. A comparable effect is produced by the introduction of a swelling agent (Fig. 9.8), which similarly retards or eliminates the crystallization.

Pure homogeneous strain

A more crucial test of the validity of the statistical theory is provided by observations on the more general state of strain, i.e. the pure homogeneous strain, involving principal stresses t_1 and t_2 corresponding to two independently variable extension ratios λ_1 and λ_2. The method of producing this state of strain in a rubber sheet, and the corresponding stress–strain relations, are considered fully in Chapter 10; the birefringence measurements, which made use of the experimental arrangement shown diagrammatically in Fig. 9.4, involved only a minor addition to enable a beam of light to be transmitted normally through the biaxially strained sheet. Fig. 9.9 shows that the birefringence in the plane of the sheet $(n_1 - n_2)$ is

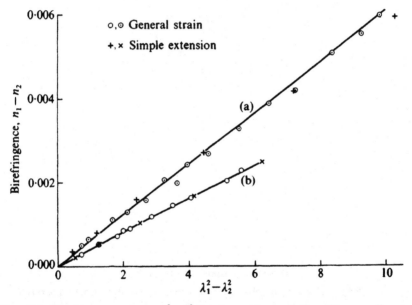

FIG. 9.9. Birefringence versus $\lambda_1^2 - \lambda_2^2$ for dry rubber and for rubber swollen to $v_2 = 0.525$ in medicinal paraffin.

directly proportional to the difference of the *squares* of the principal extension ratios (Treloar 1947), in accordance with the theoretical relation (9.14). The relation between the birefringence and the difference of principal stresses (Fig. 9.10) is also linear (eqn (9.16)), and since the points corresponding to simple extension fall on the

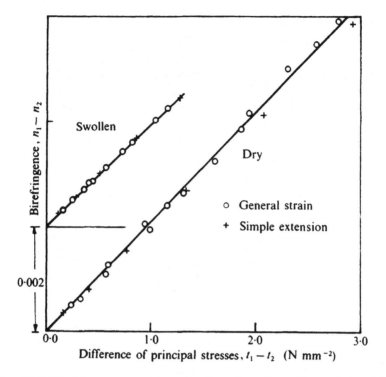

FIG. 9.10. Birefringence versus $t_1 - t_2$ for dry and swollen rubber as in Fig. 9.9.

same line as the points for the more general strain, the value of the stress-optical coefficient is the same for both types of strain.

It is to be noted that while the incorporation of a swelling liquid leads to a reduction in the slope of the birefringence–strain relation (Fig. 9.9), as required by the theory (eqn (9.17)), the slope of the birefringence–stress relation in Fig. 9.10 (as also in Fig. 9.8) is substantially unaffected. However, later work by Gent (1969), referred to later, suggests that this result may not apply generally to all types of swelling liquid.

Absolute refractive indices

The theoretical treatment implies that the total (mean) polarizability of the network, being independent of the *orientation* of the polarizable units, should be independent of strain; the same should apply also to the mean refractive index n_0, which is equal to $\frac{1}{3}(n_1 + n_2 + n_3)$. For the case of simple extension, for which $n_2 = n_3$, it follows that the relation

$$n_1 - n_0 = -2(n_2 - n_0) \qquad (9.24)$$

should apply, i.e. that the change in n_1, the refractive index corresponding to the direction of extension, should be in the opposite sense to the change in the refractive index for the transverse direction, and of twice the amount. Fig. 9.11, reproduced from the

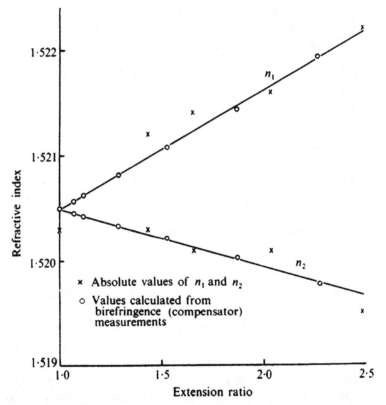

FIG. 9.11. Absolute values of n_1 and n_2 compared with values calculated from birefringence (compensator) measurements. (Saunders 1956.)

work of Saunders (1956), confirms this relationship. The crosses represent direct refractometer measurements of n_1 and n_2 independently, while the circles are calculated in accordance with (9.24) from the values of birefringence obtained with a Babinet compensator.

9.8. The effect of the degree of cross-linking

The results reported in the preceding section may be regarded as typical, and are sufficient to establish the validity of the statistical theory for the interpretation of the form of the dependence of birefringence on stress and on strain. This confirmation is the more significant in that the theoretical relations were derived independently of, and indeed prior to, their experimental verification.

We turn now to the more detailed question of the significance of the numerical value of the stress-optical coefficient and its interpretation in relation to the structure of the molecular chain. As already noted, eqn (9.12) implies that this quantity (C) should be independent of degree of cross-linking; it is therefore a characteristic parameter of the particular molecular structure.

Early work by Thibodeau and McPherson (1934) on sulphur-vulcanized rubbers revealed a marked dependence of C on degree of cross-linking, as represented by the percentage of combined sulphur. This, however, could have arisen from a spurious effect of side reactions with sulphur leading to chemical modification of the chain structure and hence of the chain polarizability. A more appropriate value in relation to the optical properties of the rubber molecule itself would be that obtained by extrapolation of the Thibodeau and McPherson data to zero sulphur content. By such extrapolation Saunders (1950) obtained a value for C of $1 \cdot 96 \times 10^{-3} \, \text{mm}^2 \, \text{N}^{-1}$; this compares with the figure of $2 \cdot 05 \times 10^{-3} \, \text{mm}^2 \, \text{N}^{-1}$ which he obtained from direct measurements on natural rubber vulcanized by means of an organic peroxide, which introduces cross-links without in any way modifying the structure of the chain. Later more extensive investigations (Saunders 1956), using both peroxide and radiation methods of cross-linking, showed that except for the highest degree of cross-linking the values of stress-optical coefficient were substantially independent of degree of cross-linking and also of the method of cross-linking (Table 9.1); these experiments yielded the mean value $C = 1 \cdot 97 \times 10^{-9} \, \text{m}^2 \, \text{N}^{-1}$. The degree of cross-linking was represented by the value of M_c, calculated from either (9.25a) or (9.25b) below,

$$\rho RT / M_c = 2C_1, \tag{9.25a}$$

$$\rho RT / M_c = 2(C_1 + C_2), \tag{9.25b}$$

where C_1 and C_2 are the constants in the Mooney equation.

TABLE 9.1

Dependence of stress-optical coefficient for natural rubber on chain molecular weight M_c (Saunders 1956)

Cross-linking process	M_c (from eqn (9.25a))	M_c (from eqn (9.25b))	Stress-optical coefficient $(10^{-9}\, m^2\, N^{-1})$
Peroxide	2 500	2 150	1·83
	5 890	4 210	1·95
	8 084	5 270	1·98
	14 160	7 400	2·00
Radiation	12 380	7 334	1·95
	20 160	13 360	2·01
	45 860	15 290	2·09

Further experiments by Saunders (1956) on peroxide-vulcanized gutta-percha, (*trans*-polyisoprene), at a temperature of 85 °C, i.e. above its crystal melting point, yielded similar results; these are given in Table 9.2.

TABLE 9.2

Dependence of stress-optical coefficient for gutta-percha on chain molecular weight M_c at 85 °C (Saunders 1956)

M_c	Stress-optical coefficient $(10^{-9}\, m^2\, N^{-1})$
3 740	3·11
4 100	3·08
6 060	3·06
6 710	3·09
13 370	3·06
35 640	3·09
Mean	3·06

Unvulcanized rubber

The above observations give strong support to the theoretical deduction that the stress-optical coefficient is a fundamental molecular parameter, whose value does not depend on the degree of cross-linking of the network. This conclusion receives further

confirmation from an experiment of Saunders (1956) on unvulcanized natural rubber. In this experiment a sample of smoked
sheet was subjected to successively increased loading, and the
birefringence and strain were measured immediately after the
application of the load and also just prior to the application of the
next additional load, during which time appreciable further extension or creep occurred. Notwithstanding this time dependence, the
resulting relation between birefringences and stress for both sets of
data was represented by a single straight line (Fig. 9.12), whose
slope (namely, $1 \cdot 92 \times 10^{-9} \, \text{m}^2 \, \text{N}^{-1}$) is within 3 per cent of the mean
value obtained for the cross-linked rubber.

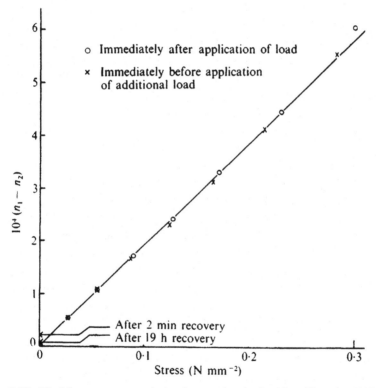

FIG. 9.12. Birefringence–stress relation for unvulcanized rubber. (Saunders 1956.)

This important result may be readily interpreted if it is assumed
that the rate of breakdown of the entanglements or 'physical'
cross-linkages between chains is not too rapid, so that the chains are

able to assume their quasi-equilibrium conformations under the influence of the stress acting at any instant. Under these conditions the birefringence, which is of course a function of the molecular orientation rather than the over-all macroscopic strain, will be directly related to the stress in the same way as in a permanently cross-linked network.

9.9. Polyethylene

Particular interest is attached to the polyethylene chain on account of the simplicity of its molecular constitution and its frequent use as a model for the treatment of the statistical properties of long-chain molecules. It might therefore be expected to offer advantages from the standpoint of the molecular interpretation of photoelastic properties.

This expectation is not wholly realized, owing to the fact that, unlike typical rubbery molecules, the polyethylene chain is subject to rather strong intramolecular interactions which interfere with the freedom of rotation about bonds, i.e. with its flexibility. However, its study does bring out a number of aspects of the problem which, though less conspicuous in other systems, are nevertheless of fundamental importance.

Polythene is a crystalline polymer, hence for photoelastic studies it must be examined at temperatures above the crystal melting point. The degree of crystallinity is widely variable. In ordinary 'low-density' polythene the chains contain a number of side-branches; these have the effect of limiting the degree of crystallinity to the range 50 per cent to 70 per cent approximately. Polythenes prepared by Ziegler-type catalysts, on the other hand, are substantially linear, and have degrees of crystallinity ranging up to about 90 per cent, with correspondingly high densities. It is also possible to prepare 'polymethylenes' of linear form by suitable chemical processes; these are also of the high-density type.

Saunders (1956) studied a series of low-density polythenes which had been subjected to varying degrees of cross-linking by irradiation in an atomic pile. From Fig. 9.13 it is seen that their stress-strain properties, which were measured at 130 °C, conform approximately with the Gaussian network theory, giving linear plots of stress against $\lambda^2 - 1/\lambda$, where λ is the extension ratio. Linear plots of birefringence versus stress were also obtained, but the slope (i.e. the stress-optical coefficient C) decreased substan-

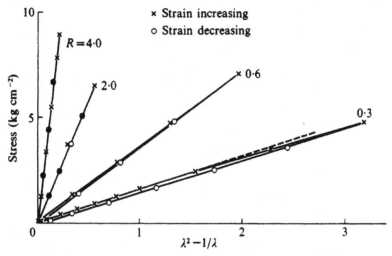

FIG. 9.13. Stress as function of $\lambda^2 - 1/\lambda$ for cross-linked polythenes with various radiation doses R. Initial molecular weight 32 000. (Saunders 1956.)

tially with increasing degree of cross-linking (Fig. 9.14). The collected data in Fig. 9.15, which include both low-density polythenes and high-density polymethylenes, appear to indicate a fairly definite relationship between C and chain molecular weight M_c.

It was originally suggested that this effect could be explained in terms of the non-Gaussian network theory of photoelasticity, but later work by Saunders, Lightfoot, and Parsons (1968), based on a different choice of bond polarizability values (see below), has cast doubt on the acceptability of this hypothesis. On the other hand, Gent and Vickroy (1967), who worked with polythenes cross-linked by irradiation at a temperature above the melting point of the crystals (i.e. cross-linked while in the amorphous state), found no significant variation of stress-optical coefficient with degree of cross-linking over the range of chain molecular weight covered ($M_c \simeq 2000$–10 000), as is seen from Fig. 9.16. There was, however, some slight evidence that for comparison samples cross-linked in the crystalline state, as in Saunders' experiments, the values of C tended to fall at high degrees of cross-linking. Gent and Vickroy concluded that the effect obtained by Saunders was probably due to the chains in the crystalline polythene being in an extended or non-random state at the instant of cross-linking; this could lead to a higher value of modulus through the factor $\overline{r_0^2}$, the mean-square

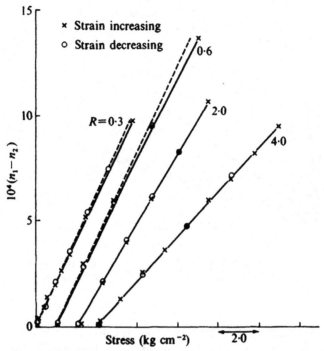

FIG. 9.14. Birefringence–stress relations for polythenes as in Fig. 9.13. (Saunders 1956.)

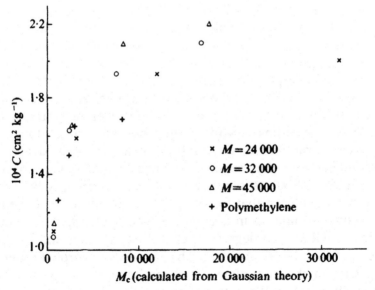

FIG. 9.15. Dependence of stress-optical coefficient C on M_c for polythenes of different initial molecular weight M. (From data of Saunders 1956.)

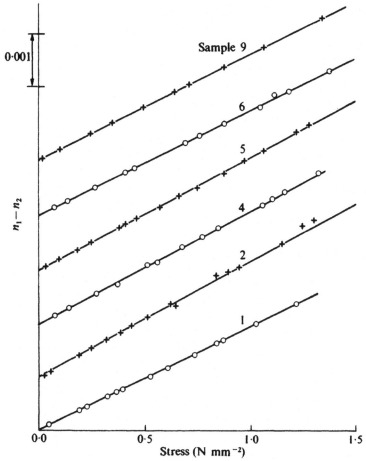

FIG. 9.16. Birefringence–stress relations for cross-linked polythenes at 169 °C. Nos 1, 2, 4, 5, and 6 cross-linked at 145 °C; No. 9 cross-linked at 40 °C. (Gent and Vickroy 1967.)

chain length in the unstrained network used in the calculation of the entropy (eqn (4.7), p. 63). Cf. also Chapter 13, eqn (13.3), p. 272).

This conclusion, however, is not confirmed by Saunders *et al.* (1968), who showed that the dependence of C on degree of cross-linking was precisely similar for samples cross-linked in either the crystalline or the amorphous state; they also criticized Gent and Vickroy's interpretation of their data and maintained that there was in fact no inconsistency between the two sets of experiments. In any case the inference drawn by Gent and Vickroy that non-random cross-linking would lead to a lower value of stress-optical coefficient is incorrect, for on the basis of the Gaussian network theory an identical factor $\overline{r^2}$ occurs in the equations for the polarizabilities (eqn (9.7)) as for the stress, leaving the value of the stress-optical coefficient C unchanged. (This factor does not appear explicitly in the elementary theory, in which it is assumed that $\overline{r^2} = nl^2$.) Indeed, the birefringence should be regarded as more specifically related to the stress than to the strain; this is borne out by the observations on stress-optical coefficient during creep, referred to in § 9.8.

It may also be noted that a comparable dependence of stress-optical coefficient on M_c was obtained by Mills and Saunders (1968) for silicone rubbers. Since these materials are non-crystalline the suggested explanation of Gent and Vickroy would not apply in this case. No satisfactory explanation of the phenomenon has up to the present been suggested.

9.10. Optical properties of the monomer unit

We have already seen that by the use of the Gaussian network theory the value of the quantity $\alpha_1 - \alpha_2$, the optical anisotropy of the random link in the hypothetical random chain, may be obtained directly from the experimental value of the stress-optical coefficient, through eqn (9.12). It has also been emphasized that the latter quantity has a characteristic value for any given polymer, which theoretically (and with certain exceptions also practically) is independent of chain length or degree of cross-linking. The question now arises of the relation between the optical anisotropy of the random link and the bond structure of the actual molecular chain. If we can calculate independently the optical anisotropy of the repeating unit in the chain, a comparison of this with the optical anisotropy of the random link should enable us to determine the number of monomer units in the equivalent random link. This in effect would provide a measure of the statistical properties of the chain, i.e. the ratio of the r.m.s. length to the fully extended length.

For the calculation of the anisotropy of the monomer unit we may apply the principle of the additivity of bond polarizabilities, discussed by Denbigh (1940). According to this principle the contribution of a given bond to the total molecular polarizability is independent of the chemical compound in which the bond is found. Values of the two principal polarizabilities (longitudinal and transverse) calculated on this basis from data for the Kerr constant and for the depolarization of scattered light (both relating to molecules in the vapour state) have been given by Denbigh. An alternative method of derivation, based on an analysis of the measured values of the principal refractive indices of a number of crystalline materials of known structure, has been put forward by Bunn and Daubeny (1954). The values for the longitudinal (b_l) and transverse (b_t) polarizabilities for the three bonds of main concern in the present context are given in Table 9.3.

TABLE 9.3
Values of longitudinal and transverse bond polarizabilities (10^{-31} m^3)

Bond	Bunn and Daubeny		Denbigh	
	b_l	b_t	b_l	b_t
C—C	10·0	2·5	18·8	0·2
C=C	29·0	10·7	28·6	10·6
C—H	8·2	6·0	7·9	5·8

It is unfortunate that for the most important bond, namely, the C—C single bond, the agreement between the two sets of figures is very poor, and therefore the choice to be made will have a profound affect on the resulting calculations. Originally Saunders (1957) chose the data of Bunn and Daubeny, but in their later work Saunders *et al.* (1968), taking into account the discussion of this problem by Volkenstein (1963), as well as internal evidence from the analysis of the temperature dependence of $\alpha_1 - \alpha_2$ (see below), concluded that the Denbigh values were probably the more reliable.

The calculation of the polarizabilities of the monomer unit in terms of the individual bond polarizabilities involves the equations of transformation (9.3). The polarizabilities required are the three polarizabilities corresponding respectively to the direction of the axis of the fully extended chain and to the two transverse directions; in general, these directions do not coincide with the principal axes of the polarizability ellipsoid. For the *cis*-isoprene repeating unit (rubber), shown in Fig. 9.17, Ox is taken parallel to the chain axis, while Oy and Oz represent the two transverse directions, the plane xOy corresponding to the plane of the main valence bonds. In general we are concerned only with the longitudinal polarizability b_l and the *mean* transverse polarizability b_t for the unit. These are given by

$$b_l = b_{xx}; \qquad b_t = \tfrac{1}{2}(b_{yy} + b_{zz}). \tag{9.26}$$

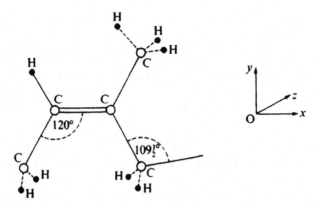

FIG. 9.17. The isoprene unit in the rubber chain (schematic). Bonds represented by dotted lines do not lie in the plane of the paper.

Calculated values of the three independent polarizabilities, and of the mean anisotropy $b_l - b_t$ for natural rubber, gutta-percha, *cis*-polybutadiene and polyethylene are given in Table 9.4.

TABLE 9.4

Directional polarizabilities and optical anisotropies of monomer units $(10^{-31}\ m^3)$ (Morgan and Treloar 1972)

Polymer	b_{xx}	b_{yy}	b_{zz}	$b_l - b_t$
Natural rubber	111·71	103·63	65·51	27·14
Gutta-percha	112·90	103·55	65·51	28·37
cis-polybutadiene	93·52	77·19	50·90	29·47
Polyethylene	24·21	19·29	14·60	7·21

9.11. The equivalent random link

The ratio of the experimentally derived anisotropy of the postulated random link in the Kuhn–Grün theory to the calculated anisotropy of the monomer unit, i.e.

$$q = \frac{\alpha_1 - \alpha_2}{b_l - b_t},\tag{9.27}$$

provides a basis for estimating the number of monomer units which are statistically equivalent to one random link. Values of this quantity for the materials listed in Table 9.4 are given in Table 9.5.

TABLE 9.5

Values of 'equivalent random link' (Morgan and Treloar 1972)

Monomer unit	$b_1 - b_t$ (10^{-31} m^3)	$\alpha_1 - \alpha_2$ (10^{-31} m^3)	Monomer units per random link q	C—C bonds per random link	q (free rotation model)
C_5H_8 *cis* (natural rubber)	27·1$_4$	47·1§	1·73	5·19	0·77
C_5H_8 *trans* (gutta-percha)	28·3$_7$	96·2$_5$†	3·39	10·17	1·25
‡C_4H_6 96 per cent *cis* (polybutadiene)	29·2$_3$	77·7$_1$	2·66	7·94	0·77
CH_2 (polyethylene)	7·2$_1$	133·4¶	18·5	18·5	3·00

†Extrapolated from data at higher temperatures by Saunders (1956).
‡ Calculated on basis 96 per cent *cis*; 4 per cent *trans*.
§ Saunders (1956, 1957).
¶ Extrapolated from data at higher temperatures by Saunders *et al.* (1968).

In view of the fact that rotation is possible only about single C—C bonds in the chain backbone, a more appropriate basis for comparison of the different structures would be in terms of the number of C—C bonds per random link; these also are given in the table. In addition 'theoretical' values of q based on the assumption of completely free rotation about single bonds are included for comparison.

It is seen that for each of the molecules examined the estimated value of q is considerably higher than that calculated on a free-rotation basis, as is of course to be expected, since the latter neglects the effects of steric hindrances, or energy barriers restricting rotation. Of more interest is the conclusion that for the *trans*-isoprene structure the value of q is nearly twice that for *cis*-isoprene, implying that the former has a more extended mean chain conformation. The greater tendency of gutta-percha to crystallize, as evidenced by its melting point (65 °C)—higher than that for natural rubber, which is normally amorphous at room temperature—is in keeping with this observation. Removal of the CH_3 side-group from *cis*-polyisoprene to produce *cis*-polybutadiene might also be expected to favour a more extended conformation by facilitating closer alignment or packing of the chains; the data also lend support to this supposition. This polymer, however, due to the lack of stereo-regularity (head-to-tail; tail-to-head) resulting from the polymerization process, is not capable of crystallization. But the most significant result is that for polyethylene, for which the number of C—C bonds per random link (namely, 18.5) is well above that for any of the other structures given. This high 'stiffness' is in harmony with the great facility with which this polymer crystallizes, and the comparatively high crystal melting point (110–30 °C). In part these features are a reflection of the basic geometry of the polyethylene chain, as shown by the comparatively large value of the length of the random link calculated on the basis of free rotation, but the more important contribution must surely be associated with energetic interactions favouring the extended rather than the folded conformations.

It should perhaps be noted that an even higher figure for polyethylene (48·5 CH_2 units per random link) would be obtained on substitution of the bond polarizability data of Bunn and Daubeny in place of those of Denbigh (Saunders 1957).

Comparison with other methods

The assignment of a value of q—the number of repeating units in the equivalent random link—determines the number of equivalent random links in a chain of specified length (molecular weight), and hence effectively defines its statistical properties over the whole range of extension (r values). This aspect has already been fully discussed in Chapter 6. Figures for the equivalent random link for natural rubber derived from the non-Gaussian network theory were there shown to vary considerably, according to the assumptions introduced into the calculations. The simple statistical theory, as treated by James and Guth, yielded the value $q = 1·63$, in fair agreement with the photoelastic estimate (1.73). The Mooney modification of the non-Gaussian statistical theory yielded the value $q = 1·5$ according to the (corrected) Mullins analysis, and $q = 4·3$ on the basis of the more complete Morris analysis. In view of the wide spread in these estimates, the most that can be said is that the photoelastic value is certainly acceptable.

9.12. The effect of swelling on stress-optical coefficient

Reference was made in § 9.7 to the observation that the stress-optical coefficient C appeared to be not significantly affected by swelling, in accordance with the prediction of the statistical theory.

Later work by Gent (1969), however, on both natural rubber and gutta-percha swollen with a wide range of organic liquids, shows wide variations in stress-optical coefficient, and suggests that the writer's earlier result may have been fortuitous. Gent's results for natural rubber are given in Table 9.6 in terms of $\alpha_1 - \alpha_2$, the optical anisotropy of the random link, the values of which are arranged in regularly increasing order.

TABLE 9.6

Random-link anisotropy $(\alpha_1 - \alpha_2)$ *derived from stress-optical co-efficient for swollen rubber* (Gent 1969)

Swelling liquid	Liquid refractive index	Swelling ratio v_2^{-1}	$b_1 - b_2$ for liquid $(10^{-30} m^3)$	$\alpha_1 - \alpha_2$ $(10^{-30} m^3)$
Benzene	1·498	6·0	1·8	3·85
CS$_2$	1·624	6·1	9·6	4·05
CCl$_4$	1·458	7·8	0·0	4·15
Toluene	1·494	6·7	3·0	4·45
None	(1·520)	—	—	4·80
n-Decane	1·406	4·6	—	4·95
p-Xylene	1·492	6·1	1·7	5·40
p-Dichlorobenzene	1·538	7·3	8·3	6·05
Biphenyl	1·604	4·3	3·8	6·45

It so happens that for both the solvents used by the author the effect was very slight. In the case of toluene (Fig. 9.8) this is directly confirmed by Gent's data. For medicinal paraffin (Fig. 9.10) a comparable figure is not available, but the result for n-decane given in the table suggests that in this case also any effect should be small.

The statistical theory assumes that the directional polarizabilities of the individual links in the long-chain molecule are completely unaffected by the presence of neighbouring molecules, whether of the polymer or of the swelling liquid. The justification for this assumption is by no means obvious, since the polarization of any given unit in the assembly of polarizable units by the electric field could well modify the internal field and thus affect the polarization of neighbouring units. If this were the cause of the effects observed by Gent the change in $\alpha_1 - \alpha_2$ would be expected to depend either on the mean polarizability (i.e. on the refractive index n) or on the anisotropy of polarizability $(b_1 - b_2)$ of the swelling liquid. The results, however, show no correlation between $\alpha_1 - \alpha_2$ and either of these quantities (Table 9.6); it is concluded, therefore, that internal field effects are not responsible. The factor which appears to be most closely correlated with the changes in $\alpha_1 - \alpha_2$ on swelling is the *geometrical* anisotropy or axial ratio of the molecule of the swelling liquid; the liquids listed in Table 9.6 are in fact also in order of the axial ratios of their molecules. It is inferred that liquid molecules with high axial ratios tend to align themselves in a direction parallel to neighbouring polymer chains and hence to acquire a degree of orientation similar to that existing in the strained polymer network. This will automatically produce an additional contribution to the strain birefringence, assuming the anisotropy of polarizability to be roughly correlated with the geometrical anisotropy of the liquid molecule.

This reasoning led Gent to suggest that in the unswollen rubber the surrounding chains similarly contribute to the observed optical anisotropy of the random link, and that their effect is about equal to that for the swelling liquids toluene and n-decane which come nearest to dry rubber in Table 9.6. On this basis the 'true' value of $\alpha_1 - \alpha_2$ for the isolated polymer molecule should be taken not as that derived from the stress-optical coefficient for the unswollen rubber but as the value derived from the stress-optical coefficient of the rubber when swollen in a liquid whose molecule is approximately spherical in form, i.e. the lowest value of $\alpha_1 - \alpha_2$ in the table. The reduction in $\alpha_1 - \alpha_2$, and in the corresponding value of q for rubber, if this procedure were adopted would amount to approximately 20 per cent.

In the case of polythene, however, Gent and Vickroy (1967) found a reduction of $\alpha_1 - \alpha_2$ from the value $8 \cdot 5 \times 10^{-30}$ m^3 for the unswollen polymer to $3 \cdot 9 \times 10^{-30}$ m^3 on swelling in decalin, i.e. a reduction of over 50 per cent. The corresponding value of q derived from the data for the swollen polymer would be only $5 \cdot 4$ CH$_2$ units per random link (Gent 1969), compared with the figure of $18 \cdot 5$ CH$_2$ units per random link given in Table 9.5. Part of this difference, however, arises from the fact that the data in this table refer to a temperature (extrapolated) of 20 °C, whereas Gent's value is calculated for the temperature of measurement, namely 102–138 °C (see § 9.13).

9.13. Temperature dependence of optical anisotropy

The expression (9.12) for the stress-optical coefficient derived on the basis of the Gaussian network theory indicates that this coefficient should be inversely proportional to absolute temperature, provided, of course, that the remaining parameters in the equation, and in particular the optical anisotropy of the random link, $\alpha_1 - \alpha_2$, are not themselves dependent on temperature. In a real chain, however, the presence of energy barriers to internal rotation implies that the statistical length of the chain, and hence the length of the equivalent random link, are functions of temperature.

Temperature coefficients of link anisotropy $\alpha_1 - \alpha_2$ are readily obtained by measurements of stress-optical coefficients over a convenient range of temperature. In general, the calculated values of $\alpha_1 - \alpha_2$ are found to be consistent with an Arrhenius type of variation, namely,

$$\alpha_1 - \alpha_2 = A \exp(E/RT), \qquad (9.28)$$

where E is equivalent to an activation energy. The value of E may be obtained from a plot of $\ln(\alpha_1 - \alpha_2)$ against $1/T$. Data for the polymers previously considered are listed in Table 9.7. A positive

TABLE 9.7

Temperature coefficients of link anisotropy (cal mol^{-1}) (Morgan and Treloar 1972)

	$R\dfrac{d\ln(\alpha_1 - \alpha_2)}{d(1/T)}$
Natural rubber	-270
Gutta-percha	$+40$
Polybutadiene (96 per cent *cis*)	$+85$
Polyethylene	$+1090$

sign indicates that the link anisotropy, and hence the link length, *decreases* with increase of temperature. It is seen that for the rubbery polymers the temperature dependence is rather slight, but that in the case of polyethylene it is large, and in the sense that the link length decreases with increasing temperature. This is consistent with the greater 'stiffness' of the polythene chain compared with the rubbers and with its preference on energetic grounds for the extended form——properties which are also implied by the high values——of q for this material.

The more detailed interpretation of the temperature dependence of $\alpha_1 - \alpha_2$ for polythene has been discussed by Saunders *et al.* (1968) in terms of the energy barriers to internal rotation. Fig. 9.18 shows

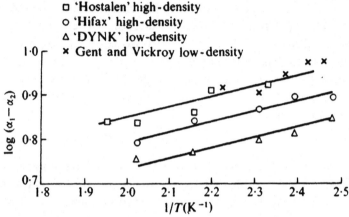

FIG. 9.18. Temperature dependence of $\log_{10}(\alpha_1 - \alpha_2)$ for various polythenes. The lines correspond to an energy difference between *trans*- and *gauche*-configurations of 1150 cal mol^{-1} on Sack's theory. (From Saunders *et al.* 1968.)

data obtained by them for the temperature dependence of three types of polythene. These data can be fitted by a theoretical treatment of the statistics of the chain given by Sack (1956) if a value of 1150 cal mol^{-1} (4830 J mol^{-1}) is taken for the energy difference between the *trans*-configuration (the state of lowest energy) and the '*gauche*'-configuration, corresponding to a rotation through 120° with respect to the *trans*-configuration.

The above conclusions regarding the temperature dependence of the random link length, and hence of the statistical length of the chain, are in qualitative agreement with independent assessments based on the thermodynamic analysis of the stress–temperature relations, discussed in Chapter 13.

10

THE GENERAL STRAIN: PHENOMENOLOGICAL THEORY

10.1. Introduction

IN previous chapters attention has been given to the deviations from the Gaussian statistical theory, which in the case of simple extension may be represented to a close approximation by the two-constant formula of Mooney. It has also been noted that in the case of uniaxial compression these deviations are either non-existent or very much less evident. From this it follows that the Mooney equation is not capable of providing a completely satisfactory and self-consistent basis for the representation of the general form of the strain-energy function for all types of strain.

In order to obtain a more accurate mathematical formulation of the general properties of rubber it is necessary to have recourse to a 'phenomenological' method of approach, i.e. to a method which is based not on molecular or structural concepts but on purely mathematical reasoning. The aim of such a method is essentially to find the most general (or convenient) way of *describing* the properties; it is not in itself concerned with their *explanation* or interpretation in the molecular or physical sense. However, before any molecular explanation is sought it is obviously highly desirable that the properties of the material shall be presented in the most general and complete form so as to avoid false deductions based on inadequate data.

The various forms of phenomenological theory which have been developed vary widely both in their generality and in their degree of sophistication, from those which do little more than describe in mathematical terms a particular stress–strain relation, e.g. simple extension, to those which attempt to relate all types of strain on the basis of one or more fundamental postulates. Our interest will be primarily in the latter type, and in the present chapter we shall examine in some detail two such theories, the original theory of Mooney as applied to the general strain problem, and the later development due to Rivlin. A more comprehensive survey of the whole field of phenomenological theory, which will include some more recent developments, will be given in the subsequent chapter.

10.2. The theory of Mooney

The earliest significant phenomenological theory of large elastic deformations, which has played a dominant part in all later work in the field, is that of Mooney (1940). It is to be noted that this theory appeared some years before the Gaussian statistical theory, so that its evolution obviously bore no relation to the deviations from the latter theory as such; Mooney was concerned simply with the problem of developing a theory of large elastic deformations *ab initio*.

Actually, Mooney's theory was developed in two forms—a special and a general. Almost all applications have been concerned with the special form, and it is to this that the present discussion will be restricted.

Mooney's theory, in the above-mentioned form, is based on the following assumptions:

(1) that the rubber is incompressible, and isotropic in the unstrained state;

(2) that Hooke's law is obeyed in simple shear, or in a simple shear superimposed in a plane transverse to a prior uniaxial extension or compression. (The more general theory is based on an arbitrary (non-linear) stress–strain relation in shear).

Of these two assumptions, the first, as has been noted in connection with the statistical theory, is in very close agreement with experiment. Regarding the second, Hooke's law is known to apply *approximately* to simple shear up to moderately large strains (cf. Fig. 5.9, p. 93; the assumption that it also applies in any plane transverse to a prior uniaxial extension or compression (i.e. to an isotropic plane) is, however, far more general.

On the basis of these assumptions Mooney derived, by purely mathematical arguments involving considerations of symmetry, the strain-energy function

$$W = C_1(\lambda_1^2 + \lambda_2^2 + \lambda_3^2 - 3) + C_2(1/\lambda_1^2 + 1/\lambda_2^2 + 1/\lambda_3^2 - 3), \quad (10.1)$$

which contains the two elastic constants C_1 and C_2. It is seen that the first term in this expression corresponds identically to the form derived from the Gaussian network theory (eqn (4.9a), p. 64), with

$$2C_1 \equiv G = NkT; \tag{10.2}$$

the Gaussian theory is thus the particular case of the Mooney theory corresponding to $C_2 = 0$. For the case of simple shear the principal extension ratios are given by $\lambda_3 = 1/\lambda_1$, $\lambda_2 = 1$; eqn (10.1) then gives

$$W = (C_1 + C_2)(\lambda_1^2 + 1/\lambda_1^2 - 2) = (C_1 + C_2)\gamma^2, \tag{10.3}$$

the equivalent shear strain γ being equal to $\lambda_1 - 1/\lambda_1$. The shear stress is

$$t_\gamma = dW/d\gamma = 2(C_1 + C_2)\gamma, \tag{10.4}$$

which corresponds to Hooke's law, the modulus of rigidity (shear modulus) being $2(C_1 + C_2)$.

It may easily be shown that a linear shear stress versus shear strain relation applies equally to a shear strain in a plane perpendicular to a prior uniaxial extension or compression.

For a simple extension (or uniaxial compression) we have $\lambda_2^2 = \lambda_3^2 = 1/\lambda_1$, and eqn (10.1) becomes

$$W = C_1(\lambda_1^2 + 2/\lambda_1 - 3) + C_2(1/\lambda_1^2 + 2\lambda_1 - 3) \tag{10.5}$$

Differentiation with respect to λ_1 then gives the force f per unit unstrained area in the form previously quoted (eqn (5.23), p. 95), namely (dropping the subscript),

$$f = 2(\lambda - 1/\lambda^2)(C_1 + C_2/\lambda). \tag{10.6}$$

The corresponding true stress t is given by

$$t = 2(\lambda^2 - 1/\lambda)(C_1 + C_2/\lambda). \tag{10.6a}$$

From the work of Gumbrell, Mullins, and Rivlin (1953), considered in Chapter 5, we have seen that the experimental data for simple extension are consistent with eqn (10.6), and that the ratio C_2/C_1 may vary from about 0·3 to 1·0. On the other hand, the results for equi-biaxial extension, which is equivalent to uniaxial compression, require that C_2 should be approximately zero, in accordance with the statistical theory (cf. Fig. 5.6, p. 90). This inconsistency is more clearly evident from a Mooney plot of the data of Rivlin and Saunders (1951) for simple extension and uniaxial compression (Fig. 10.1). In the extension region $(1/\lambda < 1)$ the Mooney line corresponds to a ratio C_2/C_1 of about 0·8, but in the

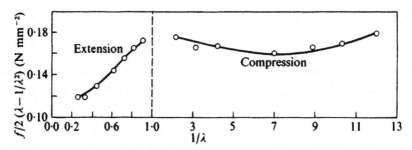

FIG. 10.1. Data for simple extension and uniaxial compression. (Rivlin and Saunders 1951.) (Note change of scale at $\lambda = 1$.)

compression region $(1/\lambda > 1)$ we have $C_2 \simeq 0$. It is worth noting that if the Mooney line representing the extension data were extrapolated into the compression region the force at $1/\lambda = 12$ would attain a value of more than *ten times* the observed value at this point. Thus, if we consider the extension and compression data together, it is clear that the Mooney equation is no improvement over the statistical theory; indeed it is very much *less* effective. Since uniaxial compression is only one of the various possible types of strain to be considered, it must follow *a fortiori* that as a general representation of the form of the strain-energy function the Mooney equation is quite inadequate.

A further and more fundamental difficulty appears on examination of the curve for simple shear (Fig. 5.9, p. 93) derived from experimental data for the equivalent pure shear. This shows deviations from linearity of the same kind, and of a similar order of magnitude, to the deviations from the statistical theory in the case of simple extension (Fig. 5.4, p. 87). The deviations from the basic postulate of the Mooney theory (Hooke's law in simple shear) are thus of the same order as the deviations in simple extension which this theory is being used to interpret. Any such interpretation is therefore logically inadmissible.

The true significance of these inconsistencies will become apparent when we come to examine the experimental data for the general biaxial strain and the interpretation of these data on the basis of the more general theory of Rivlin, which we now proceed to examine.

10.3. Rivlin's formulation

For a complete specification of the mechanical properties of a rubber it is necessary to define the form of the strain-energy function W for a pure homogeneous strain of the most general type (Fig. 4.1, p. 62). Rivlin (1948) considered from a purely mathematical standpoint the problem of the most general possible form which the strain-energy function could assume. At first sight it might appear that this could be represented by *any* arbitrarily chosen function of whatever form; closer examination, however, shows that this is not so, and that considerations of logical consistency, together with certain implied assumptions of the problem, themselves introduce restrictions on the forms which this function may assume.

Rivlin took as his basic assumptions that the material is incompressible and that it is isotropic in the unstrained state. The condition for isotropy requires that the function W shall be *symmetrical* with respect to the three principal extension ratios λ_1, λ_2, and λ_3. Furthermore, since the strain energy is unaltered by a change of sign of two of the λ_i, corresponding to a rotation of the body through 180°, Rivlin argued that the strain-energy function must depend only on the *even* powers of the λ_i. The three simplest possible even-powered functions which satisfy these requirements are the following:

$$I_1 = \lambda_1^2 + \lambda_2^2 + \lambda_3^2,$$

$$I_2 = \lambda_1^2\lambda_2^2 + \lambda_2^2\lambda_3^2 + \lambda_3^2\lambda_1^2, \qquad (10.7)$$

$$I_3 = \lambda_1^2\lambda_2^2\lambda_3^2.$$

These three expressions, being independent of the particular choice of coordinate axes, are termed *strain invariants*. Any more complex even-powered function of the λ_i can always be expressed in terms of these three basic forms.†

The condition for incompressibility or constancy of volume during deformation introduces the further relation

$$I_3 = \lambda_1\lambda_2\lambda_3 = 1, \qquad (10.8)$$

---

† In the second edition of this book (though not in the first) it was implied (p. 155) that W was *necessarily* an even-powered function of the λ_i. This is incorrect. The λ_i referred to in Rivlin's analysis are related to the equation for the strain ellipsoid, which when referred to its principal axes takes the form $x^2/\lambda_1^2 + y^2/\lambda_2^2 + z^2/\lambda_3^2 = 1$, and hence yields a solution in terms of the λ_i^2 (cf. Rivlin (1948, eqns 3.3)). This solution does not discriminate between positive and negative roots. If, however, the λ_i are re-defined, in conformity with physical reality, as the principal semi-axes of the strain ellipsoid, which are essentially positive quantities, the restriction to an even-powered function is evaded.

which enables the remaining two strain invariants to be written in the form

$$I_1 = \lambda_1^2 + \lambda_2^2 + \lambda_3^2,$$
$$I_2 = 1/\lambda_1^2 + 1/\lambda_2^2 + 1/\lambda_3^2. \tag{10.9}$$

The quantities I_1 and I_2 may be regarded as two independent variables which are determined by the three extension ratios (of which, for an incompressible material, only two are independent). The strain-energy function for an incompressible isotropic elastic material may therefore be expressed as the sum of a series of terms, thus

$$W = \sum_{i=0, j=0}^{\infty} C_{ij}(I_1 - 3)^i (I_2 - 3)^j, \tag{10.10}$$

involving powers of $(I_1 - 3)$ and $(I_2 - 3)$. These quantities are chosen in preference to I_1 and I_2 in order that W shall vanish automatically at zero strain ($I_1 = I_2 = 3$); for the same reason $C_{00} = 0$.

In the absence of any prior knowledge derived either from observation or from a structural or molecular theory there is no way of selecting a set of terms, or of assigning values to the constants in (10.10) to represent the behaviour of an actual material. However, from considerations of mathematical simplicity alone it might reasonably be expected that a small number of terms, corresponding to the lowest members of the series, would predominate. The first of such terms, obtained by putting $i = 1$, $j = 0$, namely,

$$W = C_{10}(I_1 - 3) = C_{10}(\lambda_1^2 + \lambda_2^2 + \lambda_3^2 - 3), \tag{10.11}$$

represents the form of strain-energy function derived from the Gaussian network theory (eqn (4.9a)). This is therefore one of the two simplest possible forms which could have been chosen. The other, namely,

$$W = C_{01}(I_2 - 3) = C_{01}(1/\lambda_1^2 + 1/\lambda_2^2 + 1/\lambda_3^2 - 3), \tag{10.12}$$

has no obvious application to rubber. However, the combination of these two expressions yields the Mooney equation, i.e.

$$W = C_{10}(I_1 - 3) + C_{01}(I_2 - 3), \tag{10.13}$$

which is equivalent to (10.1) above. The Mooney equation is thus seen to be *the most general first-order relationship in I_1 and I_2*.

It may be helpful to visualize the general strain-energy function $W(I_1, I_2)$ as a curved surface in space, when plotted along the vertical axis of a rectangular coordinate system in which the two horizontal axes are $I_1 - 3$ and $I_2 - 3$. The most general *first-order* function of I_1 and I_2 will be represented by a plane. If also $W = 0$ when $I_1 - 3 = I_2 - 3 = 0$, this plane (OABC in Fig. 10.2) passes

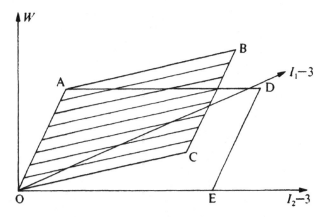

FIG. 10.2. Representation of the strain energy W as a function of the two strain invariants. OABC = Mooney theory; OADE = statistical theory.

through the origin. This corresponds to the Mooney form of strain-energy function. The particular case $\partial W/\partial I_2 = 0$, corresponding to the statistical theory, is represented by a plane (OADE) which contains the $I_2 - 3$ axis.

It can immediately be understood that any continuous curved surface (whose gradients are also continuous) can be approximated, over a sufficiently small area, by its tangent plane, which in the present case is represented by the equation

$$W = A + C_1(I_1 - 3) + C_2(I_2 - 3), \qquad (10.14)$$

which differs from the Mooney equation (10.13) only in the presence of an additive constant A. Since the stresses are determined only by the changes in W, this additive constant is of no physical significance. Hence, so long as we restrict ourselves to a limited range of the variables, the Mooney equation *necessarily* provides a fair approximation to the behaviour of the material.

It follows from this that great care must be exercised in drawing general conclusions from the apparent agreement with the Mooney theory obtained experimentally. Experiments involving any particular type of strain, e.g. simple extension or simple shear, provide too restricted a basis for the derivation of the true form of W. In such simple types of strain the variables I_1 and I_2 are related to each other, as is shown for certain particular cases in Fig. 10.3. Any such simple strain traces out a single line on the strain-energy surface. It is only by considering all types of strain (i.e. the general biaxial strain) and covering as wide a range of strain as possible, that the true picture can be assessed with any hope of reliability.

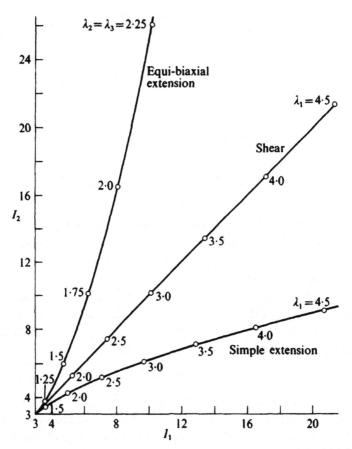

FIG. 10.3. Relation between I_1 and I_2 for particular types of strain.

10.4. Pure homogeneous strain

The principal stresses corresponding to a pure homogeneous strain of the most general type may be derived by an argument similar to that given on p. 65. These involve the partial derivatives of the strain-energy function with respect to the independent variables I_1 and I_2, and are expressed by the equation (Rivlin 1948)

$$t_i = 2\left(\lambda_i^2 \frac{\partial W}{\partial I_1} - \frac{1}{\lambda_i^2} \frac{\partial W}{\partial I_2}\right) + p \qquad (i = 1, 2, 3), \qquad (10.15)$$

where p is an arbitrary hydrostatic stress. By subtraction this may be eliminated to give the three principal stress differences, i.e.

$$t_1 - t_2 = 2(\lambda_1^2 - \lambda_2^2)\{\partial W/\partial I_1 + \lambda_3^2(\partial W/\partial I_2)\},$$

$$t_2 - t_3 = 2(\lambda_2^2 - \lambda_3^2)\{\partial W/\partial I_1 + \lambda_1^2(\partial W/\partial I_2)\}, \qquad (10.16)$$

$$t_3 - t_1 = 2(\lambda_3^2 - \lambda_1^2)\{\partial W/\partial I_1 + \lambda_2^2(\partial W/\partial I_2)\}.$$

Simple elongation

For the particular case of simple extension we obtain, on putting $\lambda_2^2 = \lambda_3^2 = \lambda^{-1}$, $t_2 = t_3 = 0$,

$$t = 2\left(\lambda^2 - \frac{1}{\lambda}\right)\left(\frac{\partial W}{\partial I_1} + \frac{1}{\lambda} \frac{\partial W}{\partial I_2}\right), \qquad (10.17)$$

where t is the tensile stress and λ the extension ratio. On putting $\partial W/\partial I_1 = C_1$, $\partial W/\partial I_2 = C_2$, in accordance with the Mooney form of W, this yields the result given previously (eqn (10.6a)).

Simple shear

Simple shear is not a *pure* strain, and the relation between shear stress t_γ and shear strain γ is therefore not obtainable directly from eqns (10.16). However, substitution of the values $\lambda_3 = 1/\lambda_1$, $\lambda_2 = 1$ into eqns (10.9) gives

$$I_1 - 3 = I_2 - 3 = \lambda_1^2 + 1/\lambda_1^2 - 2 = \gamma^2, \qquad (10.18)$$

since $\gamma = \lambda_1 - 1/\lambda_1$. The corresponding shear stress has been shown by Rivlin (1948) to be given by

$$t_\gamma = 2\left(\frac{\partial W}{\partial I_1} + \frac{\partial W}{\partial I_2}\right)\gamma. \qquad (10.19)$$

For the case when $\partial W/\partial I_1$ and $\partial W/\partial I_2$ are both constants (C_1 and C_2, respectively) this reduces to

$$t_\gamma = 2(C_1 + C_2)\gamma \qquad (10.20)$$

in agreement with eqn (10.4) above.

10.5. The general strain: early experiments

To devise an experiment which will allow the application of two independently variable strains in two perpendicular directions with the simultaneous measurement of the stresses is not altogether simple. The technique devised by the author (Treloar 1948), though not ideal, seems to meet the requirements sufficiently well, and enables moderately large strains (i.e. up to $\lambda = 3$) to be attained. In this method a suitably marked sheet of rubber having the dimensions shown in Fig. 10.4(a) is extended in two directions at right

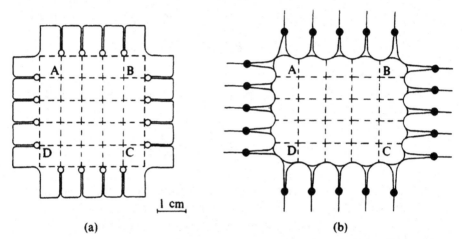

(a)   (b)

FIG. 10.4. Biaxial extension test-piece. (a) Unstrained; (b) strained (smaller scale). The middle area ABCD is in state of pure homogeneous strain.

angles by means of strings attached to a number of projecting lugs on its perimeter. Of the five strings attached to one side, the middle three were loaded with equal weights while the two outermost were independently adjusted so as to secure a uniform distribution of strain in the middle region (rectangle ABCD). The principal extension ratios λ_1 and λ_2 in the plane of the sheet were directly measured, the third extension ratio λ_3 being then calculable from the incompressibility condition (eqn (10.8)). The principal stresses t_1 and t_2 in the plane of the sheet were obtained from the forces applied to the sides AB and CD of the rectangular block ABCD. The third principal stress t_3 was of course zero.

Fig. 10.5 shows a plot of the difference of principal stresses $t_1 - t_2$ in the plane of the sheet against $\lambda_1^2 - \lambda_2^2$ for a sulphur-vulcanized natural rubber compound (Treloar 1948). Data are included also

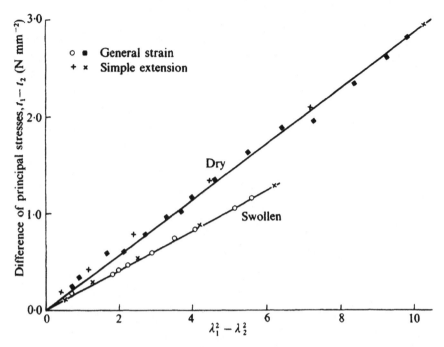

FIG. 10.5. Biaxial extension. Difference of principal stresses $t_1 - t_2$ plotted against $\lambda_1^2 - \lambda_2^2$ for rubber in unswollen and swollen ($v_2 = 0.525$) states.

for the same rubber swollen to nearly twice its original volume ($v_2 = 0.525$) in medicinal paraffin. In these experiments the values of λ_1 and λ_2 ranged from about 0.7 to 3.2 for the dry rubber and from about 0.7 to 2.5 for the swollen material. In both states the linear relation (4.18) (p. 67) required by the statistical theory was approximately satisfied, though significant differences are detectable in the case of the dry rubber. A different picture emerged, however, when t_1 and t_2 were plotted separately against $\lambda_1^2 - \lambda_3^2$ and $\lambda_2^2 - \lambda_3^2$, respectively. According to eqns (4.18) we should have

$$t_1 = t_1 - t_3 = G(\lambda_1^2 - \lambda_3^2),$$
$$t_2 = t_2 - t_3 = G(\lambda_2^2 - \lambda_3^2). \tag{10.21}$$

The data for the unswollen rubber, plotted on this basis, are shown in Fig. 10.6, from which it is apparent that eqns (4.18) are not satisfied.

These data were not analysed in terms of $\partial W/\partial I_1$ and $\partial W/\partial I_2$; they were, however, compared with the Mooney equation, for

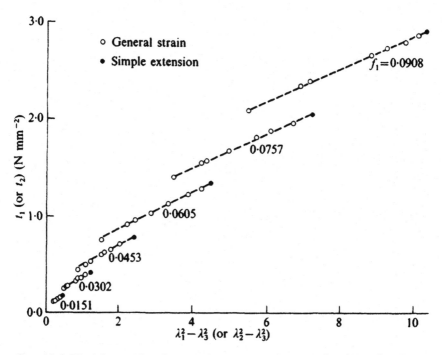

FIG. 10.6. Biaxial extension. Stress t_1 (or t_2) plotted against $\lambda_1^2 - \lambda_3^2$ (or $\lambda_2^2 - \lambda_3^2$) for unswollen rubber.

which $\partial W/\partial I_1 = C_1$ and $\partial W/\partial I_2 = C_2$, so that eqns (10.16) become

$$t_1 - t_3 = 2C_1(\lambda_1^2 - \lambda_3^2)\{1 + (C_2/C_1)\lambda_2^2\}, \quad \text{etc.} \qquad (10.22)$$

According to this, a plot of t_1 against $(\lambda_1^2 - \lambda_3^2)\{1 + (C_2/C_1)\lambda_2^2\}$, with a suitable choice of the ratio C_2/C_1, should yield a straight line. This was found to be the case for the swollen rubber, with $C_2/C_1 = 0.10$, but for the dry rubber the best result obtainable, using $C_2/C_1 = 0.05$, yielded a curve rather than a straight line, indicating that C_1 could not be treated as a constant, i.e. the Mooney equation was not capable of providing an accurate representation of the data (Fig. 10.7).

The reason for the apparent rather close agreement with the statistical theory in the plots of $t_1 - t_2$ against $\lambda_1^2 - \lambda_2^2$ shown in Fig. 10.5 will appear later.

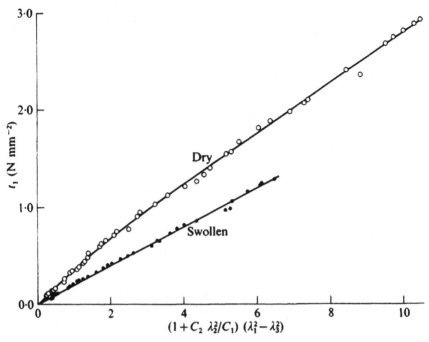

FIG. 10.7. Biaxial extension. Stress t_1 (or t_2) plotted on basis of eqn (10.22). (a) Unswollen rubber, using $C_2/C_1 = 0.05$; (b) swollen rubber, using $C_2/C_1 = 0.10$.

10.6. The experiments of Rivlin and Saunders

The data obtained from the above experiments in which λ_1 and λ_2 were not separately controlled were not well adapted to determining with any degree of precision the form of the strain-energy function in terms of I_1 and I_2, or of its derivations $\partial W/\partial I_1$ and $\partial W/\partial I_2$ which occur in the general stress–strain relations (10.16). An important advance was introduced by Rivlin and Saunders (1951), who adopted the more logical procedure of choosing the conjugate values of λ_1 and λ_2 in the biaxial strain experiment in such a way that in any given test one of the two strain invariants I_1 and I_2 was held constant while the other was varied. To achieve this end they introduced a set of helical springs to replace the dead-weight loading used by the author; the strain in either direction could therefore be continuously varied at will and the corresponding stresses determined from the spring extensions.

Since $t_3 = 0$, the values of t_1 and t_2 corresponding to any given values of λ_1 and λ_2 (and hence of I_1 and I_2) are readily derivable from eqns (10.16). The explicit solution is as below,

$$2\frac{\partial W}{\partial I_1} = \left(\frac{\lambda_1^2 t_1}{\lambda_1^2 - \lambda_3^2} - \frac{\lambda_2^2 t_2}{\lambda_2^2 - \lambda_3^2}\right)(\lambda_1^2 - \lambda_2^2)^{-1},$$

$$2\frac{\partial W}{\partial I_2} = -\left(\frac{t_1}{\lambda_1^2 - \lambda_3^2} - \frac{t_2}{\lambda_2^2 - \lambda_3^2}\right)(\lambda_1^2 - \lambda_2^2)^{-1}. \tag{10.23}$$

The resulting dependence of $\partial W/\partial I_1$ and $\partial W/\partial I_2$ on both I_1 and I_2 obtained in this way is shown in Fig. 10.8. The continuous lines in

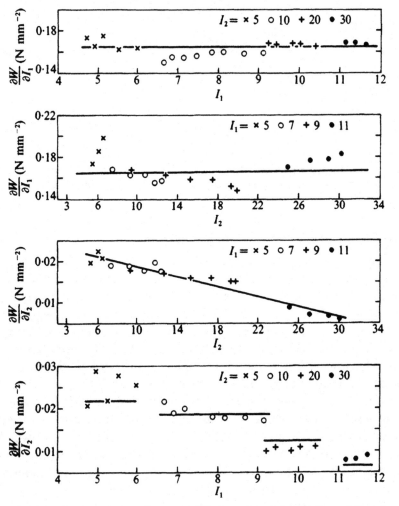

FIG. 10.8. Dependence of $\partial W/\partial I_1$ and $\partial W/\partial I_2$ on I_1 and I_2. (Rivlin and Saunders 1951.)

the separate graphs are drawn in such a way as to be self-consistent, on the assumption that $\partial W/\partial I_1$ is a constant (independent of I_1 and I_2), while $\partial W/\partial I_2$ is a function of I_2 but is independent of I_1 (for a given value of I_2). It is seen that the major term, for all states of strain, is $\partial W/\partial I_1$, and that $\partial W/\partial I_2$ varies from about $\frac{1}{8}$ of $\partial W/\partial I_1$ at small values of I_2 to about $\frac{1}{30}$ of $\partial W/\partial I_1$ at the highest values employed. It was therefore tentatively suggested that the strain-energy function has the general form

$$W = C_1(I_1 - 3) + \Phi(I_2 - 3), \qquad (10.24)$$

in which Φ is a function whose slope diminishes continuously with increasing I_2.

The accuracy of these results is limited by the accuracy of the original data, which are themselves limited not so much by the accuracy of the measurements of stress and strain in themselves, but rather by the inherent lack of complete reproducibility (reversibility) of the rubber. The occurrence of $(\lambda_1^2 - \lambda_2^2)^{-1}$ as a factor in eqns (10.23) magnifies the experimental errors in the resultant values of $\partial W/\partial I_1$ and $\partial W/\partial I_2$, particularly in the region where $\lambda_1 - \lambda_2$ is small, rendering the results in this region rather unreliable. This accounts for the larger scatter of the points in the lower ranges of I_1 and I_2 shown in Fig. 10.8.

10.7. Interpretation of Mooney plots

The conclusions arrived at by Rivlin and Saunders enable the apparent applicability of the Mooney equation in the case of simple extension to be more critically interpreted. In this state of strain, as pointed out in § 10.3, the variables I_1 and I_2 are not independent (cf. Fig. 10.3), and it is therefore not possible in this case to apply eqns (10.23) so as to obtain the values of $\partial W/\partial I_1$ and $\partial W/\partial I_2$ at each particular value of the strain. The constants C_1 and C_2 derived from a Mooney plot cannot therefore be identified with the quantities $\partial W/\partial I_1$ and $\partial W/\partial I_2$ in eqns (10.16) unless it is known (from independent experiments) that both $\partial W/\partial I_1$ and $\partial W/\partial I_2$ are independent of strain. But the experiments of Rivlin and Saunders show conclusively that $\partial W/\partial I_2$ is strongly dependent on strain. Thus, while the relations $\partial W/\partial I_1 = C_1$, $\partial W/\partial I_2 = C_2$ *necessitate* a linear Mooney plot in the case of simple extension, the converse is not true, and the apparent consistency of the data with the Mooney equation could equally well be attributed to a variation of $\partial W/\partial I_1$ or

$\partial W/\partial I_2$ (or both) with strain. That this is in fact the correct explanation is proved not only by the work of Rivlin and Saunders but also by later experiments referred to in the next chapter.

An explicit warning against the näive identification of the Mooney constants derived from simple extension data with the partial derivatives $\partial W/\partial I_1$ and $\partial W/\partial I_2$ in the general strain-energy function has been given by Rivlin and Saunders (1951) but seems to have been generally disregarded. Such an identification, as they point out, is quite inconsistent with their own general strain data, and also with results which they present in relation to shear and torsion. It also gives a misleading impression of the magnitude of the deviations from the statistical theory. In a particular experiment which they quote, the ratio C_2/C_1 derived from a Mooney plot of simple extension data was $0\cdot8$, whereas the *maximum* ratio of $\partial W/\partial I_2$ to $\partial W/\partial I_1$ derived from their biaxial strain data was about $0\cdot125$.

In the light of the general strain data of Rivlin and Saunders, the close agreement with the statistical theory observed in the case of uniaxial compression can be readily understood. The explanation lies in the particular range of the variables I_1 and I_2 covered by the different types of strain. For a simple extension with $\lambda = 5$, for example, the values of $I_1 - 3$ and $I_2 - 3$ are $22\cdot4$ and $7\cdot04$, respectively; for a uniaxial compression with $\lambda_1 = \frac{1}{5}$, however, these figures are reversed, i.e. $I_1 - 3 = 7\cdot04$, $I_2 - 3 = 22\cdot4$. The higher values of I_2 in uniaxial compression result in much smaller values of $\partial W/\partial I_2$ in this type of strain.

Again, the absence of obvious deviations from the statistical theory in the plot of $t_1 - t_2$, the principal stress difference in the plane of the sheet, against $\lambda_1^2 - \lambda_2^2$, for the general biaxial strain (Fig. 10.5) can be understood on the basis of the general stress–strain relations (10.16). In this case both λ_1 and λ_2 are greater than unity, and since $\lambda_3 = 1/\lambda_1\lambda_2$ we shall have, in general, $\lambda_3^2 \ll 1$. Consequently the effect of the term $\lambda_3^2(\partial W/\partial I_2)$ in the first of eqns (10.16) will rapidly diminish with increasing strain. On the other hand, plots of $t_1 - t_3$ or $t_2 - t_3$ (Fig. 10.6) will involve the terms $\lambda_2^2(\partial W/\partial I_2)$ or $\lambda_1^2(\partial W/\partial I_2)$, which are not small compared with $\partial W/\partial I_1$; these will therefore show significant deviations from the statistical theory.

Finally, it should be pointed out that the general validity of Rivlin and Saunders' conclusions, as embodied in eqn (10.24), must be subject to some uncertainty, on account of the experimental limitations, referred to above, and also, of course, to possible variations between different rubber vulcanizates. Later work, referred to in the next chapter, does in fact suggest that some minor modifications may be desirable at least in particular cases. Such modifications are however rather marginal in character and do not alter the broad conclusions originally established by Rivlin and Saunders concerning the form of the deviations from the statistical theory.

10.8. Molecular significance of deviations from statistical theory

It has already been indicated that no generally accepted molecular interpretation of the deviations from the statistical theory discussed above has yet been advanced. Possible interpretations are usually presented as explanations of the Mooney C_2 term, as deduced from experiments in simple extension, but from the evidence presented in the previous section the reader will appreciate that this definition of the problem is too narrow, and may be positively misleading. The real problem is to explain the general form of the strain-energy function, and in particular the presence of a $\partial W/\partial I_2$ term which decreases with increasing I_2.

Among possible molecular explanations of the observed phenomena the following are perhaps the most plausible:

(1) non-Gaussian chain or network statistics;
(2) Internal energy effects;
(3) chain entanglements;
(4) irreversible effects;
(5) non-random packing effects.

Non-Gaussian chain or network statistics

It has been suggested by Isihara, Hashitsume, and Tatibana (1951) that the non-Gaussian network theory could give rise to a deviation of the type represented by a Mooney C_2 term. Without taking into account the quantitative values of higher-order terms in relation to the C_2 term, this in itself is not very significant, and indeed Thomas (1955) has claimed that their theory yields the wrong sign for the C_2 term.

Internal energy effects

It would indeed be reasonable to expect the effect of energetic interactions between chains to give rise to significant deviations from the simple statistical theory. This point of view has been put forward by Wang and Guth (1952). However, the characteristic variation of C_2 (in simple extension) with degree of swelling, discussed in § 10.3, and its independence of both the type of rubber and the nature of the swelling liquid, would appear to preclude such an explanation. It would also be necessary to explain the absence of deviations from the theory in the case of uniaxial compression.

Krigbaum and Koneko (1962) have treated the statistics of long-chain molecules on the basis of a lattice model, taking into account energetic interactions within the chain. A non-Gaussian network theory incorporating such molecules gives a suggestion of a deviation of the kind observed in simple extension.

A suggestion by Thomas (1955) that an empirical term, representing an internal attractive force, be introduced into the expression for the free energy of the single chain has been shown to lead to general stress–strain relations showing rather close agreement with the experimental data of Rivlin and Saunders. This is one of the few attempts to relate the *general* strain properties to a structural characteristic, though it leaves open the question of the possible origin of such an additional term.

Chain entanglements

A strong case can be made for thinking that the explanation for the deviations from the statistical theory must be connected with some general topological features of the network. Similar types of deviation to that observed with natural rubber are found in a wide variety of cross-linked networks, e.g. butyl and silicone rubbers (Ciferri and Flory 1959), hydrofluorocarbon (Viton), rubbers (Roe and Krigbaum 1963), and polythene (Gent and Vickroy 1967). A likely explanation might be in terms of chain entanglements, which may act as partial cross-links (cf. Moore and Watson, 1956, p. 168). The mathematical difficulties involved in any attempt to work out a quantitative theory on these lines are, however, formidable. An approach by Prager and Frisch (1967), based on an extremely simplified specification of an 'entanglement' leads to potentially interesting conclusions, but the authors themselves are reluctant to predict the effect of entanglements in a real network.

Irreversible effects

The case for attributing the deviations to irreversible effects (hysteresis) has been strongly argued by Ciferri and Flory (1959), but the supporting evidence is inconclusive. These authors also fail to consider that any such general explanation should apply equally to all types of strain (e.g. should include uniaxial compression, for which the deviations are non-existent). It is interesting to note, however, that Kraus and Moczvgemba (1964) found that networks prepared from carboxy-terminated polybutadienes, which contain no 'loose ends' of the type encountered in conventional vulcanizates, and which should therefore experience less obstruction in the attainment of equilibrium, showed almost perfect agreement with the statistical theory. A similar effect is obtained by the cross-linking of natural rubbers in solution (Price *et al.* 1970). On the other hand, Krigbaum and Roe (1965), in a careful review of the evidence on the subject, point out that certain rubbers still show significant deviations from the statistical theory even when precautions are taken to avoid all irreversible effects.

Non-random packing effects

The frequently made suggestion that non-random packing of chains could modify the expression for the entropy of deformation in such a way as to lead to the observed deviations is plausible, though as in the case of entanglements, difficult to quantify. The interested reader is referred to the work of Di Marzio (1962) on this subject, and also to a recent paper by Schwarz (1973). The latter author considers the orientation of links in the chain to be modified by that of neighbouring links in other chains; as a result the chains of lowest M_c values (and hence of highest link orientation) will tend to increase the orientation of surrounding links. The treatment appears to give a plausible explanation of the observations in extension and compression represented in Fig. 10.1.

Conclusion

In contrast to the original spectacular success of the statistical theory, the failure to secure any very significant understanding of the relatively rather small deviations from the theory, despite repeated attempts over a period of 30 years, is disappointing. In a sense, perhaps, there are too many available explanations, and too few critical data by which they may be tested. The importance of the study of the form of the general strain-energy function is that it provides a much more detailed basis for the critical examination of theoretical ideas than can be obtained from the consideration of any single type of strain in isolation.

11

ALTERNATIVE FORMS OF STRAIN-ENERGY FUNCTION

11.1. Survey of alternative proposals

In this chapter some of the more recent developments bearing on the question of the form of the strain-energy function for rubber will be examined. The aim of these developments is essentially to find the most appropriate general method of characterizing the properties of a rubber for engineering applications or similar purposes, any possible interpretation in molecular terms of such methods being either irrelevant or purely incidental.

A number of authors have attempted to expand the Rivlin type of formulation, as represented by eqn (10.10) (p. 216), so as to include higher-order terms in $I_1 - 3$ and $I_2 - 3$. For reasons which were fully discussed in the last chapter a critical evaluation of such formulae requires the consideration of types of strain other than simple extension, which in some cases is the only type of strain taken into account by the authors concerned. Ideally, for a complete evaluation, general biaxial strain data covering the widest possible range of strain should be used, but if such data are not available a fair assessment can usually be made on the basis of certain widely different types of strain, e.g. simple extension and equi-biaxial extension (uniaxial compression). For purely practical reasons these may even be advantageous, particularly if the interest is in the high-strain (non-Gaussian) region, since they usually cover a wider range of deformation than it has yet proved possible to attain in the general biaxial strain test used by Rivlin and Saunders, which is limited to principal extension ratios of about 3.

In the present context little would be gained by a detailed presentation of the various formulae which have been proposed, and we shall content ourselves with a reference to a small number of examples by way of illustration. For a more detailed review the reader is referred to papers by Hart-Smith (1966) and Alexander (1968). Our main interest will be in those developments which involve distinctive features of general interest, particularly where these depart from the original formulation of Rivlin. In this connection, the theories proposed by Ogden (1972) and by Valanis and Landel (1967) will be expounded in some detail.

Among the multi-term Rivlin proposals may be mentioned that of Isihara, Hashitsume, and Tatibana (1951) which as indicated in the previous chapter is based on their development of the non-Gaussian network theory. This contained a term in $(I_1-3)^2$, in addition to the first-order terms in I_1 and I_2, i.e.

$$W = C_{10}(I_1-3) + C_{01}(I_2-3) + C_{20}(I_1-3)^2. \qquad (11.1)$$

Alexander (1968) has shown that this formula gives inconsistent results when applied to the author's data (Treloar 1944) for uniaxial extension and equibiaxial extension (Figs 5.4 (p. 87 and 5.5 p. 89). The four-term expression

$$W = C_{10}(I_1-3) + C_{01}(I_1-3) + C_{20}(I_1-3)^2 + C_{30}(I_1-3)^3 \qquad (11.2)$$

put forward by Biderman (1958) is, according to Alexander, open to a similar criticism. More elaborate formulae of a similar type have been derived by Tschoegl (1971), who showed that the complete force–extension curve for a carbon-reinforced natural rubber vulcanizate in simple extension could be accurately represented by the equation

$$W = C_{10}(I_1-3) + C_{01}(I_2-3) + C_{11}(I_1-3)(I_2-3), \qquad (11.3)$$

which contains a product term in addition to the two Mooney terms. For a butadiene–styrene 'pure gum' rubber the best fit was obtained with the formula

$$W = C_{10}(I_1-3) + C_{01}(I_2-3) + C_{22}(I_1-3)^2(I_2-3)^2. \qquad (11.3a)$$

Curve-fitting procedures of this kind may clearly be carried out to any desired degree of accuracy. The elaboration, however, is misleading for the reasons emphasized in the previous chapter, namely, that in the case of simple extension to which these results refer the I_1 and I_2 are not independent variables, but are both functions of the single variable λ. Thus all the information which can properly be deduced from such experiments can be represented in terms of an algebraic function of λ alone.

A more realistic formula, which accounts to a fair degree of approximation for the general strain data of Rivlin and Saunders (1951) has been put forward by Gent and Thomas (1958). This, which also yielded results substantially similar to those deduced from the theory of Thomas (1955) and referred to on p. 228, was of the form

$$W = C_1(I_1 - 3) + k \ln (I_2/3),$$ (11.4)

in which the second term replaces the unknown function in the equation (10.24) of Rivlin and Saunders. This form was chosen chiefly on the grounds of mathematical simplicity. Differentiation of (11.4) yields

$$\frac{\partial W}{\partial I_1} = C_1, \qquad \frac{\partial W}{\partial I_2} = \frac{k'}{I_2},$$ (11.5)

corresponding to a non-linear dependence of $\partial W/\partial I_2$ on I_2. The difference from the linear dependence provisionally indicated experimentally (Fig. 10.8, p. 224) is probably not significant.

In the region of high extensions the rapid increase of stress which occurs as the limiting extensibility of the network is approached cannot be satisfactorily reproduced by a small number of terms in $I_1 - 3$ and $I_2 - 3$. For this reason some authors have preferred to incorporate non-Gaussian type terms directly into the strain-energy function. Thus Hart-Smith (1966) has suggested the three-constant formula

$$\frac{\partial W}{\partial I_1} = G \exp \{k_1(I_1 - 3)^2\}; \qquad \frac{\partial W}{\partial I_2} = \frac{Gk_2}{I_2},$$ (11.6)

to explain a variety of data on the inflation of balloons (equi-biaxial extension). Using this equation he was also able to fit the general strain data of Rivlin and Saunders (1951). By a further elaboration of this type of approach, Alexander (1968) arrived at the more complicated five-parameter expression

$$W = C_1 \int \exp \{k(I_1 - 3)^2\} \, dI_1 + C_2 \ln \left(\frac{(I_2 - 3) + \gamma}{\gamma} \right) + C_3(I_2 - 3),$$ (11.7)

which was found to give good agreement with his experimental data for polychloroprene rubber in simple extension and equi-biaxial extension.

11.2. Ogden's formulation

The complexity of some of the formulae referred to above arises partly from the attempt to cover the region of very high strains, where the first-order terms in I_1 and I_2 in Rivlin's formulation are quite inadequate. However, if we are concerned with a purely formal representation of the elastic properties of a rubber, it is possible to argue that the introduction of the strain invariants I_1 and I_2 may itself be an unnecessary complication, and that a formulation in terms of principal extension ratios or other more direct measures of strain may have advantages in mathematical simplicity as well as in the not unimportant matter of ready intelligibility. We have already seen (p. 215, footnote) that the arguments advanced by Rivlin in support of the restriction of the strain-energy function to even powers of the extension ratios have no necessary basis in physical reality; the question of the representation of the properties in terms of the strain invariants I_1 and I_2 is therefore one of convenience rather than logical necessity.

An important new departure was made by Ogden (1972), who dispensed with the restriction to even powers of the extension ratios and wrote the strain-energy function for an incompressible rubber in the form of the series

$$W = \sum_n \frac{\mu_n}{\alpha_n}(\lambda_1^{\alpha_n} + \lambda_2^{\alpha_n} + \lambda_3^{\alpha_n} - 3), \tag{11.8}$$

in which the α_n may have any values, positive or negative, and are not necessarily integers, and the μ_n are constants. It will be noted that the statistical theory is a special case of this formula, with $\alpha_1 = 2$, while the Mooney equation contains two terms, corresponding to $\alpha_1 = 2$, $\alpha_2 = -2$. The principal stresses corresponding to the strain-energy function (11.8) are of the form

$$t_i = \sum_n \mu_n \lambda_i^{\alpha_n} - p \qquad (i = 1, 2, 3), \tag{11.9}$$

where p is an arbitrary hydrostatic stress. As in the statistical theory, the indeterminacy associated with the arbitrary pressure p is a consequence of the assumption of incompressibility and does not appear in the equations for the *differences* of principal stresses. These are of the form

$$t_1 - t_2 = \sum_n \mu_n(\lambda_1^{\alpha_n} - \lambda_2^{\alpha_n}). \tag{11.10}$$

Simple extension

For the case of simple extension $\lambda_2 = \lambda_3 = \lambda_1^{-\frac{1}{2}}$, $t_2 = t_3 = 0$. Hence from (11.10) the tensile stress is given by

$$t_1 - t_2 = t_1 = \sum_n \mu_n(\lambda_1^{\alpha_n} - \lambda_1^{-\alpha_n/2}). \tag{11.11}$$

The corresponding nominal stress f_1, or force per unit unstrained area, is

$$f_1 = t_1/\lambda_1 = \sum_n \mu_n(\lambda_1^{\alpha_n-1} - \lambda_1^{(-\alpha_n/2)-1}). \tag{11.11a}$$

Equi-biaxial extension

Putting $\lambda_1 = \lambda_2^{-2} = \lambda_3^{-2}$, $t_2 = t_3$, $t_1 = 0$, the stress in the plane of the sheet (t_2 or t_3) is given by

$$t_2 - t_1 = t_2 = \sum_n \mu_n(\lambda_2^{\alpha_n} - \lambda_2^{-2\alpha_n}). \tag{11.12}$$

The corresponding force per unit unstrained area then becomes

$$f_2 = t_2/\lambda_2 = \sum_n \mu_n(\lambda_2^{\alpha_n-1}\lambda_2^{-2\alpha_n-1}) \tag{11.12a}$$

Pure shear

For a pure shear produced by principal stresses t_1 and t_2, with $t_3 = 0$, we have $\lambda_3 = \lambda_1^{-1}$, $\lambda_2 = 1$, and hence

$$t_1 - t_3 = t_1 = \sum_n \mu_n(\lambda_1^{\alpha_n} - \lambda_1^{-\alpha_n}),$$

$$\tag{11.13}$$

$$t_2 - t_3 = t_2 = \sum_n \mu_n(1 - \lambda_1^{-\alpha_n}).$$

The corresponding forces per unit unstrained area are

$$f_1 = t_1/\lambda_1 = \sum_n \mu_n(\lambda_1^{\alpha_n-1}\lambda_1^{-\alpha_n-1}),$$

$$\tag{11.13a}$$

$$f_2 = t_2.$$

Application to experimental data

The utility of the Ogden formulation must be judged by its ability to represent the experimental data for rubber under all possible types of strain, preferably (though this is not a primary requirement) with the use of a relatively small number of terms. The above equations have been applied by Ogden to the author's data for

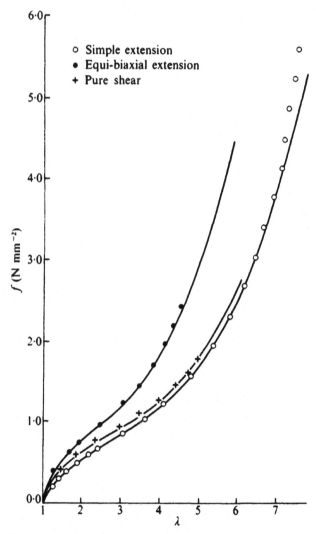

FIG. 11.1. Representation of data for simple extension, equibiaxial extension, and pure shear by Treloar (1944) on basis of eqns (11.11a), (11.12a), and (11.13a). (Ogden 1972.)

simple extension, equi-biaxial extension, and pure shear, previously given in Figs 5.4 (p. 87), 5.5 (p. 89), and 5.8 (p. 92), respectively. For simple extension and pure shear a two-term formula was found to be effective, but for the representation of all three types of strain three terms were required. The degree of agreement attained is shown in Fig. 11.1, the force f for each of the curves being that corresponding to the major axis of the strain ellipsoid, as represented by eqns (11.11a), (11.12a), and (11.13a). The agreement is very satisfactory, except at the highest values of strain $(\lambda > 7 \cdot 0)$, where a significant deviation begins to occur. The values of the six constants involved are as follows (μ in kg cm^{-2}):

$$\alpha_1 = 1 \cdot 3, \qquad \alpha_2 = 5 \cdot 0, \qquad \alpha_3 = -2 \cdot 0;$$
$$\mu_1 = 6 \cdot 3, \qquad \mu_2 = 0 \cdot 012, \qquad \mu_3 = -0 \cdot 1. \qquad (11.14)$$

Although the number of independent constants to be determined is rather large the method, considered from the purely empirical standpoint, has a certain simplicity combined with generality in that all the terms are of identical type, and no prior judgment concerning the expected form of the strain-energy function is involved. Any possible influence of the form of representation on the conclusions derived is thereby eliminated. The fact that the terms are all of the same type also has advantages in further mathematical operations involving inhomogeneous strains, since in effect all such operations are performed on a single representative term into which appropriate numerical values may then be introduced for numerical computation.

11.3. The Valanis–Landel hypothesis

Valanis and Landel (1967) have put forward the hypothesis that the strain-energy function should be capable of representation as the sum of three separate functions of the three principal extension ratios, i.e. by an equation of the form

$$W = w(\lambda_1) + w(\lambda_2) + w(\lambda_3), \qquad (11.15)$$

in which, from symmetry considerations, the separate functions $w(\lambda_i)$ are of identical form. Though there is no general physical basis for this assumption, it is pointed out by the authors that it is satisfied by the Gaussian network theory and the Mooney equation

(eqn (10.1)), and also by the simplest form of non-Gaussian network theory, i.e. that of James and Guth, in which the chains lying in
the directions of the principal axes of strain are treated as independent of each other (p. 114); it therefore has a strong *a priori*
plausibility. It may also be noted that the later theory of Ogden
(1972) referred to above is consistent with this hypothesis. Its
validity in any given case, however, must rest ultimately on experimental evidence. If it should prove to be applicable, the problem of
defining the elastic properties of a rubber over the whole range of
available strains is reduced from a complex problem in three
dimensions to the comparatively simple problem of finding the form
of the function $w(\lambda)$ of the single variable λ.

The Valanis–Landel hypothesis not only has far-reaching
theoretical implications, but has also provided a fresh stimulus to
experimental work on the most general type of strain, indicating
new approaches both to the methods of experimentation and to the
presentation of the resulting data which hold out the promise of
improved accuracy and more direct mental appreciation.

The principal stresses corresponding to the strain-energy function (11.15), in accordance with eqn (11.9), are given by

$$t_i = \lambda_i \frac{\partial W}{\partial \lambda_i} - p = \lambda_i \frac{dw}{d\lambda_i} - p. \qquad (11.16)$$

Writing $w'(\lambda)$ for $dw/d\lambda$, we have then

$$t_i = \lambda_i w'(\lambda_i) - p, \qquad (11.16a)$$

where p is an arbitrary hydrostatic stress. The principal stress
differences then become

$$t_1 - t_2 = \lambda_1 w'(\lambda_1) - \lambda_2 w'(\lambda_2), \quad \text{etc.} \qquad (11.17)$$

This equation will apply to a general biaxial strain experiment
involving the measurement of the two principal stresses t_1 and t_2 in
the directions of λ_1 and λ_2, with zero stress in the direction λ_3. It is to
be noted that if, in such an experiment, λ_2 is held constant while λ_1 is
varied, the second term on the right-hand side of eqn (11.17) is
constant; hence the variation of $\lambda_1 w'(\lambda_1)$ may be directly determined. Valanis and Landel considered in particular the case when
$\lambda_2 = 1$, corresponding to a pure shear. For this case eqn (11.17) may

be written

$$(t_1 - t_2)_{\lambda_2=1} = \lambda_1 w'(\lambda_1) - c, \qquad (11.18)$$

where c is a constant, equal to $w'(1)$. Such an experiment therefore enables the form of the function $w'(\lambda)$ to be determined, subject only to an unknown additive constant. On substituting back into eqns (11.17) to obtain the principal stress differences for any other state of strain the additive constant disappears. This derivation is restricted in its application to the range of λ_1 covered by the experiment, which of course implies values of λ greater than unity. However, the corresponding function for $\lambda < 1$ may be obtained from the equation

$$t_3 - t_2 = \lambda_3 w'(\lambda_3) - \lambda_2 w'(\lambda_2). \qquad (11.19)$$

Putting $t_3 = 0$, and remembering that $\lambda_3 = \lambda_1^{-1}$, this yields

$$-t_2 = \lambda_1^{-1} w'(\lambda_1^{-1}) - c. \qquad (11.20)$$

The function $w'(\lambda)$ is thus readily determined over a wide range of λ, both greater than and less than unity.

It is concluded from this analysis that the form of $w'(\lambda)$, and hence the stress–strain relations for any type of strain, *may be predicted on the basis of a single experiment in pure shear.*

Actually, this conclusion is not restricted to the case of pure shear ($\lambda_2 = 1$) but applies equally for any experiment at constant λ_2, whatever the value of λ_2, as can be seen by retracing the foregoing argument. In practice, by working with other values of λ_2, the range of λ over which the function $w'(\lambda)$ is determined may be extended to lower values of λ than would be possible from a pure shear experiment alone.

11.4. Experimental examination of Valanis–Landel hypothesis

The above theory has been applied by Valanis and Landel to a variety of experimental data by other authors. For this purpose previously unpublished experimental data of Becker and Landel for pure shear were used to provide the basic form of $w'(\lambda)$, and a scale factor or modulus 2μ was introduced to render the rubbers used by different authors comparable. Thus by plotting the relevant $(t_1 - t_2)/2\mu$ for any given type of strain against the corresponding $\{\lambda_1 w'(\lambda_1) - \lambda_2 w'(\lambda_2)/2\mu$ derived from the pure shear experiment, in accordance with eqn (11.17), a straight line of unit slope should

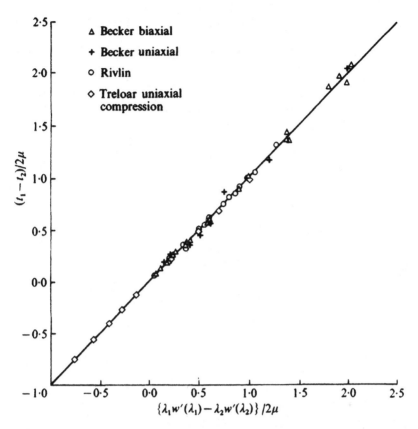

FIG. 11.2. Representation of various stress–strain data on basis of eqn (11.19), using form of $w'(\lambda)$ obtained from experiments of Becker and Landel in pure shear. (From Valanis and Landel 1967.)

result. Fig. 11.2 shows this relationship to apply with remarkable accuracy.

With regard to the form of $w'(\lambda)$, as derived from the pure shear experiment, Valanis and Landel (1967) claimed that the logarithmic relation

$$w'(\lambda) = 2\mu \ln \lambda \qquad (11.21)$$

was approximately valid over the range $\lambda = 0.35$ to $\lambda = 2.5$. This, however, was based on the incorrect assumption that the constant c in eqns (11.18) and (11.20) could without loss of generality be put equal to zero. The correct interpretation of their data would yield

$$w'(\lambda) = 2\mu \ln \lambda + c/\lambda \qquad (11.22)$$

in place of eqn (11.21).†

A more systematic examination of the applicability of eqn (11.17) has been undertaken by Obata, Kawabata, and Kawai (1970), who carried out experiments on the general homogeneous strain of a

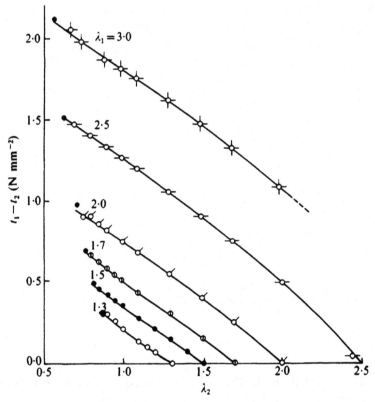

FIG. 11.3. Plot of $t_1 - t_2$ against λ_2 for different values of λ_1. (From Obata, Kawabata, and Kawai 1970.)

sheet of rubber under conditions such that one of the principal extension ratios (in this case λ_1) was held constant, while the other (λ_2) was varied. The test piece was similar to that used by Rivlin and Saunders, but the test machine employed enabled the sample to be

† The issue is further confused by the fact that the curve for $2\mu\ln \lambda$ in Fig. 2 of Valanis and Landel (1967) is incorrectly drawn (Landel, private communication).

extended continuously up to any predetermined values of λ_1 and λ_2. According to eqn (11.17) the curves representing the relation between $t_1 - t_2$ and λ_2 for different (constant) values of λ_1 are then all of the same form but are displaced vertically with respect to one another by amounts corresponding to the different values of the constant $\lambda_1 w'(\lambda_1)$. By suitable vertical displacement they should therefore all be superposable on to a single curve, e.g. that for pure shear ($\lambda_1 = 1$). Fig. 11.3 shows the data of Obata *et al.* for various values of λ_1 plotted in this way; each of these curves (except that for $\lambda_1 = 3\cdot0$) covers the range $\lambda_2 = \lambda_1^{-\frac{1}{2}}$ (simple extension) to $\lambda_2 = \lambda_1$ (equi-biaxial extension). The curves are all of similar form, and may be brought into coincidence with the curve for pure shear by appropriate vertical displacement (Fig. 11.4).

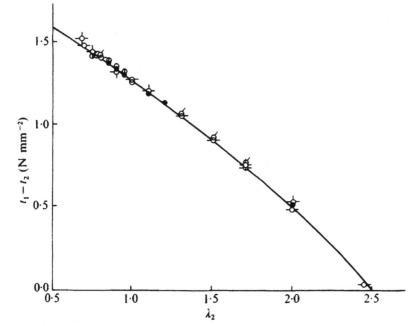

FIG. 11.4. Superposition of data shown in Fig. 11.3 on curve derived from pure shear experiment. (From Obata *et al.* 1970.)

These results demonstrate the validity of the Valanis–Landel hypothesis over the range of strain covered by the experiments, and justify the conclusion that the stress–strain relations for any type of strain may be deduced from a single experiment in pure shear.

11.5. Form of the function $w'(\lambda)$

Obata *et al.* follow Valanis and Landel in taking $c = 0$ in eqns (11.18) and (11.20) in order to calculate $w'(\lambda)$, but their results are not consistent with the logarithmic formula (11.21), or more correctly (11.22). They do not, however, suggest any alternative formula to represent their data.

Experiments similar to those of Obata *et al.* have been carried out by Jones and Treloar (1975) using the method of Rivlin and Saunders (1951). Their data for $t_1 - t_2$ (eqn (11.18)) and $-t_2$ (eqn (11.20)) for different values of the parameter λ_2 are shown in Fig. 11.5. (This figure is inverted with respect to Fig. 11.3 due to the

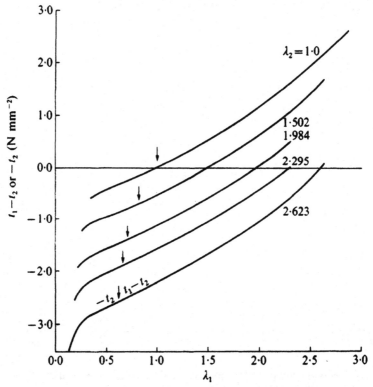

FIG. 11.5. Plot of $t_1 - t_2$ and $-t_2$ against λ_1 for different values of λ_2. (Jones and Treloar 1975.)

interchange of λ_1 and λ_2.) The individual curves were superposed by vertical displacement so as to coincide with the curve for pure shear ($\lambda_2 = 1$) at the point $\lambda_1 = 1$ (zero strain), as shown in Fig. 11.6. The

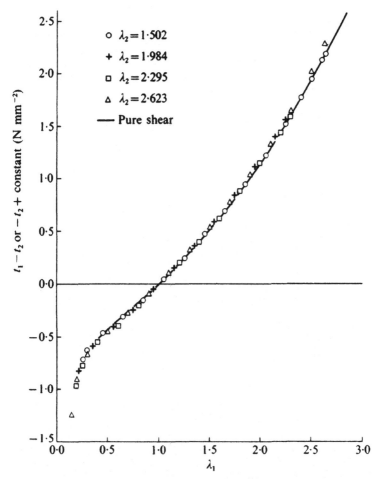

FIG. 11.6. Superposition of data shown in Fig. 11.5 for various values of λ_2, to coincide with pure shear curve at $\lambda_1 = 1$. (Jones and Treloar 1975.)

variations between the curves were in all cases within the experimental error. Since no special significance is attached to the case $\lambda_2 = 1$, the best values of the function $\lambda w'(\lambda)$ were considered to be given by the mean of all these curves, including that for $\lambda_2 = 1$; *these* values are reproduced in Fig. 11.7. The function $\lambda w'(\lambda)$ is of course indeterminate to the extent of the unknown constant c.

Application of Ogden formula

Since, as noted above, Ogden's formulation (eqn (11.8)) is consistent with the Valanis–Landel hypothesis, the analysis of the form of

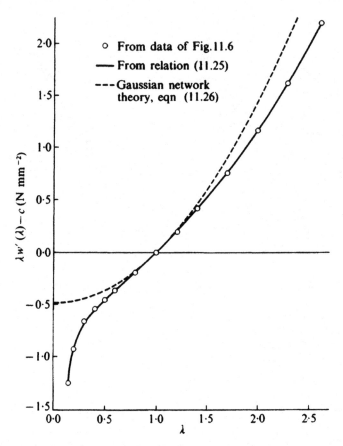

FIG. 11.7. Mean experimental values of $\lambda w'(\lambda)$ from data of Fig. 11.6 compared with relation (11.25). The broken line corresponds to the Gaussian network theory (eqn (11.26)). (Jones and Treloar 1975.)

the experimentally derived function $\lambda w'(\lambda)$ in terms of an Ogden series is an obvious possibility. Comparing eqns (11.17) and (11.19) we may write

$$t_1 - t_2 = \lambda_1 w'(\lambda_1) - \lambda_2 w'(\lambda_2) = \sum_n \mu_n (\lambda_1^{\alpha_n} - \lambda_2^{\alpha_n}). \qquad (11.23)$$

The curve for pure shear ($\lambda_2 = 1$) thus becomes

$$(t_1 - t_2)_{\lambda_2 = 1} = \lambda_1 w'(\lambda_1) - c = \sum_n \mu_n (\lambda_1^{\alpha_n} - 1). \qquad (11.24)$$

The data from Fig. 11.6 were found to be fitted by a three-term expression of the type (11.24), namely,

$$\lambda w'(\lambda) - c = 0 \cdot 69(\lambda^{1 \cdot 3} - 1) + 0 \cdot 010(\lambda^{4 \cdot 0} - 1) - 0 \cdot 0122(\lambda^{-2 \cdot 0} - 1),$$
$$(11.25)$$

in which the μ_i are given in Newtons per square millimetre. This function is represented by the continuous curve in Fig. 11.7. It will be noted that the numerical values of the α_n are very similar to those found by Ogden (eqns (11.14)).

It is interesting to compare the results obtained in this way with the Gaussian network theory. For the latter we have, from eqn (4.9a) (p. 64), $w(\lambda) = \frac{1}{2} G \lambda^2$ and hence $dw/d\lambda = G\lambda$. This gives

$$\lambda w'(\lambda) = G\lambda^2. \qquad (11.26)$$

For pure shear this yields the parabolic relation

$$t_1 - t_2 = \lambda w'(\lambda) - w'(1) = G(\lambda^2 - 1). \qquad (11.27)$$

This function is plotted in Fig. 11.7, the value of G being adjusted to give the same slope at $\lambda = 1$, i.e. the same small-strain modulus, as eqn (11.25). The comparison brings out very clearly the difference of form between the two functions, particularly (and rather surprisingly) in the region $\lambda < 1$.

The interpretation of these results must be treated with some caution. In particular, any tendency to translate the function $\lambda w'(\lambda)$ directly to the molecular system must be avoided. It would be unwise at this stage, for example, to attach any special molecular significance to the rather sharp increase in the slope of this function in the region below $\lambda = 0 \cdot 5$. A further reservation, which applies to this, as to any other phenomenological representation, is that the above conclusions regarding the applicability of the Valanis–Landel hypothesis cannot be extrapolated to values of λ outside the range covered in these experiments. Indeed, it would be expected on molecular grounds that at sufficiently high values of λ the hypothesis will cease to apply, as can be seen by considering the limiting case in which the chains are fully extended in one direction and are therefore incapable of *any* extension in the transverse directions. It is suspected that the Valanis–Landel postulate is closely connected with the property of the Gaussian network that it is equivalent to three independent sets of Gaussian chains parallel to the coordinate axes, a property which is itself derived from the Gaussian probability function for the single chain (cf. p. 48). This property of the Gaussian network, though assumed to apply also in the simple model of the non-Gaussian network (p. 113) must become increasingly inapplicable with increasing values of the strain. At present, however, this suggestion must be regarded as purely speculative.

11.6. Re-examination in terms of strain invariants

The analysis of Obata, Kawabata, and Kawai

The recent work of Obata *et al.* (1970) and of Jones and Treloar (1975) has also included a re-examination of the form of the strain-energy function in terms of the strain invariations I_1 and I_2, which leads to a slight modification of the conclusions tentatively derived from the experiments of Rivlin and Saunders (1951) on the general homogeneous strain.

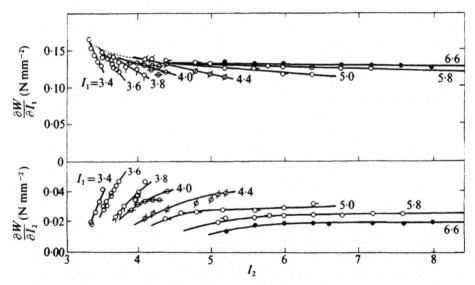

FIG. 11.8. Dependence of $\partial W/\partial I_1$ and $\partial W/\partial I_2$ on I_2, for various values of I_1. (Obata *et al.* 1970.)

Fig. 11.8 represents the results deduced from the data of Obata *et al.* by the method referred to in § 11.4 above in terms of the variation of $\partial W/\partial I_1$ and $\partial W/\partial I_2$ with I_2, at various values of I_1. Whereas Rivlin and Saunders found that $\partial W/\partial I_1$ was substantially constant, these more recent results indicate that with increasing I_2 $\partial W/\partial I_1$ at first decreases but subsequently tends towards a constant value. It is also seen that, for a given value of I_2, $\partial W/\partial I_1$ *increases* with increasing I_1. The behaviour of $\partial W/\partial I_2$ is directly opposite; this quantity *increases* with increasing I_2 (in contrast to the results of Rivlin and Saunders) and *decreases* with increasing I_1.

These apparent discrepancies from the earlier work are probably not as important as might appear at first sight. In the first place, they refer mainly to much lower strains. In the plots of Rivlin and Saunders' data (Fig. 10.8, p. 224), the *lowest* values of I_1 and I_2 were 5·0; in Fig. 11.8, the values range from about 3·3 to 8, and the most marked variations occur in the region of I_1 and I_2 *below* 5·0. Unfortunately, this is the region where the values of $\partial W/\partial I_1$ and $\partial W/\partial I_2$, as calculated from eqns (10.23), are most seriously affected by experimental errors. Secondly, there may be compensating effects between changes in $\partial W/\partial I_1$ and $\partial W/\partial I_2$ with strain, and the precise total effect on the stress can be deduced only by detailed consideration of any particular type of strain. Thirdly, it is to be noted that the ratio of $\partial W/\partial I_2$ to $\partial W/\partial I_1$ shown in Fig. 11.8 is always small, ranging from about 0·1 to 0·3. In this respect, therefore, these data are in agreement with those of Rivlin and Saunders.

In the case of simple extension, Obata *et al.* have attempted to estimate the contributions of the separate terms in $\partial W/\partial I_1$ and $\partial W/\partial I_2$ to the total force. As pointed out in the preceding chapter, these terms cannot be obtained by a direct analysis of the simple extension data. Obata *et al.* therefore estimated their values by extrapolation of their pure homogeneous strain data to the values of I_1 and I_2 corresponding to simple extension. Their results are reproduced in Fig. 11.9, which gives the estimated contributions of the two component terms to the total force, as represented by the equation

$$\frac{f}{\lambda - 1/\lambda^2} = 2\left(\frac{\partial W}{\partial I_1} + \frac{1}{\lambda}\frac{\partial W}{\partial I_2}\right) \qquad (11.28)$$

From this it is apparent that the contribution of the $\partial W/\partial I_2$ term is always small (10 per cent or less). This result differs from the conclusion drawn by Rivlin and Saunders only in that the deviations from the statistical theory are associated with changes in both $\partial W/\partial I_1$ *and* $\partial W/\partial I_2$, and are not attributed solely to changes in $\partial W/\partial I_2$.

More recently Kawabata (1973) has studied the temperature dependence of $\partial W/\partial I_1$ and $\partial W/\partial I_2$, as determined from biaxial

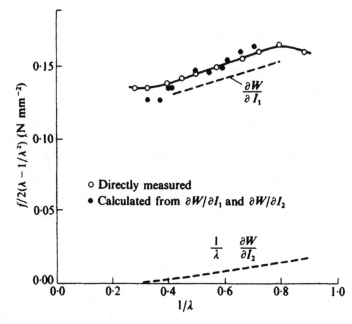

FIG. 11.9. Contributions of the $\partial W/\partial I_1$ and $\partial W/\partial I_2$ terms to the total force in simple extension. (Obata *et al.* 1970.)

strain experiments. His results (Fig. 11.10) indicate that at temperatures not too close to the transition temperature the temperature dependence of $\partial W/\partial I_1$ is in approximate accord with the simple kinetic theory, as indicated by the dotted line, while $\partial W/\partial I_2$ appears to be substantially independent of temperature over the whole range $-20\,°C$ to $+100\,°C$. These results were not dependent on the time under load. This gives some support to the view that the $\partial W/\partial I_1$ term is closely associated with the network properties, while $\partial W/\partial I_2$ has some quite different origin.

The analysis of Jones and Treloar

The advantage of the Valanis–Landel form of analysis is that the function $\lambda w'(\lambda)$, obtained by superposition of the curves corresponding to different values of λ_2, automatically smoothes out the experimental errors attached to individual points, with the consequence that the magnification of experimental errors associated with the calculation of $\partial W/\partial I_1$ and $\partial W/\partial I_2$ from isolated measurements of t_1 and t_2 through the application of eqns (10.23) no longer occurs. The calculation of these quantities is further facilitated by the use of the Ogden type of formula to represent the function $\lambda w'(\lambda)$.

Jones and Treloar (1975) used their curve for $\lambda w'(\lambda)$, as represented by the expression (11.25), to calculate values of t_1 and t_2, and hence of $\partial W/\partial I_1$ and $\partial W/\partial I_2$, for the particular pairs of values of λ_1 and λ_2, corresponding to $I_1 = $ constant and $I_2 = $ constant, respectively. Their results, reproduced in Fig. 11.11, are thus directly comparable with those of Rivlin and Saunders (Fig. 10.8, p. 224), discussed in the preceding chapter. These later results suggest that neither $\partial W/\partial I_1$ nor $\partial W/\partial I_2$ are strictly constant, but that each may

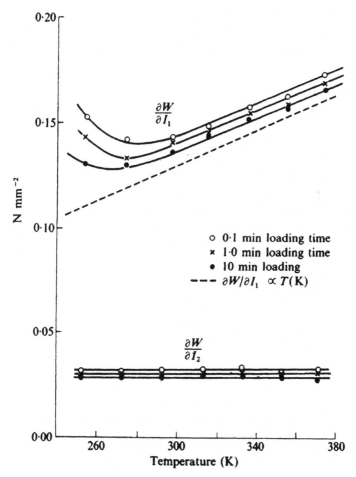

FIG. 11.10. Temperature dependence of $\partial W/\partial I_1$ and $\partial W/\partial I_2$ (at $I_1 = I_2 = 6 \cdot 0$) for natural rubber, various loading times. (Kawabata 1973.)

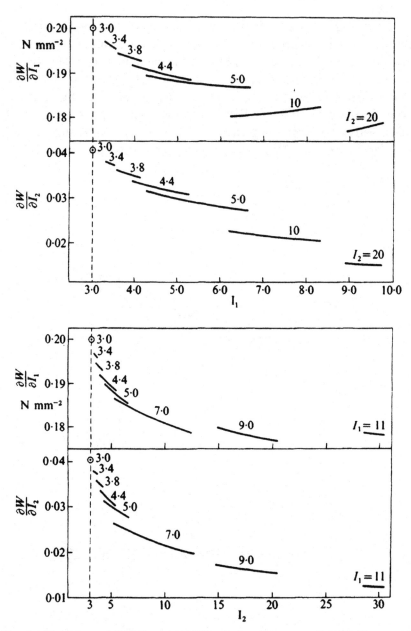

FIG. 11.11. Values of $\partial W/\partial I_1$ and $\partial W/\partial I_2$ calculated from eqn (11.25). (Jones and Treloar 1975.)

depend on both I_1 and I_2. The variations in $\partial W/\partial I_1$ are, however, *relatively* much less important than the variations in $\partial W/\partial I_2$, the extreme variations amounting only to about 10 per cent in the case of $\partial W/\partial I_1$, compared with about 70 per cent for $\partial W/\partial I_2$. There is no clear evidence that $\partial W/\partial I_2$ is any less dependent on I_1 (at constant I_2) than on I_2 (at constant I_1).

It appears from these results, as well as from those of Obata *et al.* (1970), that the conclusions arrived at by Rivlin and Saunders, and embodied in the tentative equation (10.24) for the strain-energy function, require some modification in detail. In particular, it would appear that the deviations from the statistical theory cannot be represented simply by an additional term in the strain-energy function, as they originally suggested.

12

LARGE-DEFORMATION THEORY: SHEAR AND TORSION

12.1. Introduction: components of stress

IN the preceding chapters our principal concern has been with the problem of defining the properties of a rubber under the most general type of strain, and we have seen that these properties may conveniently be represented in terms of the strain-energy W per unit volume of the material for a pure homogeneous strain. A knowledge of the form of the strain-energy function is all that is required for the treatment of more complex problems in which the strains are not homogeneous but vary from point to point in the deformed body, the analysis of such problems being purely a question of mathematics.

There are, however, a number of effects deduced from the application of large-deformation theory to practical problems such as, for example, shear and torsion, which are peculiar to large elastic deformations and are not to be anticipated on the basis of the classical small-strain theory. These effects are quite general and are not dependent for their existence on any particular form of strain-energy function. Thus, though the full treatment of the mathematical analysis of problems of this kind would not be appropriate to the present work, some indication of the important conclusions derived from the study of typical relatively simple systems, and of the way in which these conclusions are arrived at, is of considerable significance for the understanding of the phenomena of rubber elasticity.

Analysis of strain

In a pure homogeneous strain, lines which are originally parallel to the three principal extensions remain unchanged in direction in the strained state. These three directions are the directions of the principal axes of the strain ellipsoid. If, however, the state of strain is inhomogeneous. the axes of the strain ellipsoid vary both in magnitude and in direction from point to point in the strained body. In this case it is desirable to refer the strain at any point not to the local axes of strain but to a system of coordinate axes fixed in space. The strain is then defined by the partial derivatives of the displacement of a representative point with respect to the fixed coordinate system. Thus, if the point initially defined by the coordinates (x, y, z) is displaced to the new position $(x + u,\ y + v,\ z + w)$, the local strain is determined by the quantities $\partial u/\partial x, \partial u/\partial y, \partial u/\partial z, \partial v/\partial x \ldots$, etc.

Components of stress

In dealing with pure homogeneous strains we have been concerned only with the three principal stresses, which act normally on surfaces perpendicular to the three principal axes of the strain ellipsoid. In the more general case the stress acting across any plane section through the body can be resolved into three components, one normal and two tangential to the surface. For the purpose of definition it is sufficient to consider only planes normal to the coordinate axes. For the plane $x =$ constant, the components of stress may be written $t_{xx},\ t_{yx},\ t_{zx}$, of which the first is the normal component and the remaining two are the tangential components in the y and z directions, respectively. Similar considerations apply for the planes $y =$ constant and $z =$ constant. There are thus altogether nine components of stress. For the maintenance of equilibrium, however, it is readily shown that

$$t_{xy} = t_{yx}, \qquad t_{yz} = t_{zy}, \qquad t_{zx} = t_{xz},$$

and hence the number of independent components is reduced to six, of which three are normal and three tangential.

12.2. Stress components in simple shear

Geometry of strain

The relation between pure shear and simple shear was considered in Chapter 5, where it was shown that a simple shear was equivalent to a pure shear plus a rotation. For either type of strain the principal extension ratios are given by

$$\lambda_2 = 1/\lambda_1, \qquad \lambda_3 = 1. \tag{12.1}$$

For simple shear the shear strain γ is given by

$$\gamma = \lambda_1 - 1/\lambda_1. \tag{12.2}$$

Consider a sphere of unit radius deformed by simple shear in the (x, y)-plane to the form of an ellipsoid, whose section in the plane of shearing is an ellipse of semi-axes λ_1 and $1/\lambda_1$ (Fig. 12.1). The

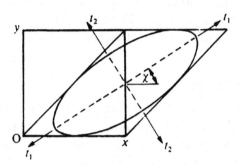

FIG. 12.1. Inclination of principal axes in simple shear.

inclination χ of the major axis to the direction of shearing, i.e. to the x-axis, is given by

$$\cot \chi = \lambda_1 \tag{12.3}$$

and is equal to 45° for small values of the shear strain ($\lambda_1 \simeq 1$), in accordance with the classical theory.

Components of stress

To examine the components of stress let us consider for simplicity that the statistical theory applies, the principal stresses being given by eqns (4.18) (p. 67). Putting $t_3 = 0$, the principal stresses t_1 and t_2 in the plane of the shear are given by

$$t_1 = G(\lambda_1^2 - 1), \qquad t_2 = G(1/\lambda_1^2 - 1). \tag{12.4}$$

To obtain the tangential and normal components of stress on any plane it is necessary to sum the components due to t_1 and t_2 using the standard formulae for the resolution of stresses. From eqn (12.3) we have $\sin^2 \chi = 1/(1+\lambda_1^2)$ and $\cos^2 \chi = \lambda_1^2/(1+\lambda_1^2)$. Making use of these relations the tangential component of stress on the plane $0y$ is given by

$$t_{yx} = t_{xy} = (t_1 - t_2) \sin \chi \cos \chi = G\left\{(\lambda_1^2 - 1) - \left(\frac{1}{\lambda_1^2} - 1\right)\right\}\frac{\lambda_1}{1+\lambda_1^2}$$

which, with (12.2), yields

$$t_{xy} = G(\lambda_1 - 1/\lambda_1) = G\gamma. \tag{12.5}$$

For the normal component of stress on the plane Ox (normal to the y-axis)

$$t_{yy} = t_1 \sin^2 \chi + t_2 \cos^2 \chi = G(\lambda_1^2 - 1)\frac{1}{1+\lambda_1^2} + G\left(\frac{1}{\lambda_1^2} - 1\right)\frac{\lambda_1^2}{1+\lambda_1^2}$$

or

$$t_{yy} = 0, \tag{12.6}$$

while for the normal component of stress on the plane Oy (normal to the x-axis)

$$t_{xx} = t_1 \cos^2 \chi + t_2 \sin^2 \chi = G\left\{(\lambda_1^2 - 1)\frac{\lambda_1^2}{1+\lambda_1^2} + \left(\frac{1}{\lambda_1^2} - 1\right)\frac{1}{1+\lambda_1^2}\right\}$$

or

$$t_{xx} = G(\lambda_1 - 1/\lambda_1)^2 = G\gamma^2. \tag{12.7}$$

Taking $t_{zz} = t_3 = 0$ there are no stresses in the z direction; the complete set of stress components thus becomes

$$t_{xx} = G\gamma^2, \qquad t_{yy} = t_{zz} = 0;$$
$$t_{xy} = t_{yx} = G\gamma; \qquad t_{yz} = t_{zy} = t_{zx} = t_{xz} = 0. \tag{12.8}$$

These components of stress are represented diagrammatically in Fig. 12.2, in which the parallelogram ABCD is deformed by simple shear to the square A'B'CD.

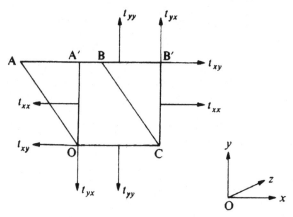

FIG. 12.2. Components of stress in simple shear. ABCD, unstrained state; A'B'CD, strained state.

From the results represented by eqns (12.8) we see that whereas the tangential components of stress t_{xy} and t_{yx} are proportional to the corresponding shear strain, in accordance with the classical theory, the normal component t_{xx} is proportional to the *square* of the strain, and has no counterpart in the classical theory. This represents an entirely new effect, associated only with the presence of *large* elastic deformations.

The component of stress t_{xx} represents a normal tensile stress acting in the direction of shearing. If this stress is not applied, there will be a contraction in this direction, with corresponding *expansions* in the lateral directions. We thus arrive at the important conclusion that *a large shear strain cannot be produced by a shear stress acting alone*, but requires in addition the application of a normal stress on at least one pair of free surfaces as in Fig. 12.2.

Although the above result has been derived on the basis of the statistical theory, similar conclusions are reached whatever the form of strain-energy function. In terms of the general strain-energy function W, Rivlin (1948) has derived the following equations for the components of stress corresponding to a simple shear in the (x, y)-plane:

$$t_{xy} = 2\gamma\{(\partial W/\partial I_1) + (\partial W/\partial I_2)\},$$

$$t_{yz} = t_{zx} = 0,$$

$$t_{xx} = 2\{(1 + \gamma^2)(\partial W/\partial I_1) - (\partial W/\partial I_2)\} + p, \qquad (12.9)$$

$$t_{yy} = 2\{(\partial W/\partial I_1) - (1 + \gamma^2)(\partial W/\partial I_2)\} + p,$$

$$t_{zz} = 2\{(\partial W/\partial I_1) - (\partial W/\partial I_2)\} + p,$$

in which p is an arbitrary constant. If we take $t_{zz} = 0$ the normal components of stress become

$$t_{xx} = 2\gamma^2(\partial W/\partial I_1), \qquad t_{yy} = -2\gamma^2(\partial W/\partial I_2), \qquad t_{zz} = 0.$$
$$(12.10)$$

Alternatively the p may be eliminated by taking *differences* between the normal stress components, i.e.

$$t_{xx} - t_{zz} = 2\gamma^2(\partial W/\partial I_1),$$
$$t_{yy} - t_{zz} = -2\gamma^2(\partial W/\partial I_2). \qquad (12.11)$$

The precise dependence of the normal stress components on the shear strain γ will depend on the form of W. In the case when both

$\partial W/\partial I_1$ and $\partial W/\partial I_2$ are independent of strain (Mooney equation) they will be proportional to γ^2; in other cases this form of dependence will require some modification. However, unless $\partial W/\partial I_1$ or $\partial W/\partial I_2$ show a very strong dependence on strain, the corresponding differences of normal stress components, represented by eqns (12.11), will be *approximately* proportional to the square of the shear strain, in contrast to the shear stress t_{xy}, which will be approximately proportional to the first power of the shear strain.

Origin of normal stress components

The existence of the normal stress components in a large simple shear is associated primarily with the lack of symmetry of the principal stresses. This is reinforced by the peculiar features of the strain geometry, as will now be shown.

In a *small* shear strain γ the principal axes of the strain ellipsoid are equal to $1 + \gamma/2$ and $1 - \gamma/2$, and are inclined at $\pm 45°$ to the direction of shearing (x-axis). The corresponding principal stresses are $+t$ (tensile) and $-t$ (compressive), as shown in Fig. 12.3, where t

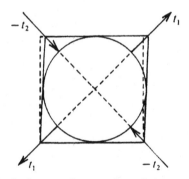

FIG. 12.3. Principal stresses in case of *small* shear strain $t_1 = -t_2 = t$.

is numerically equal to the shear stress (Love, 1934). The resultant of these stresses normal to the plane Ox is therefore

$$t_{xx} = t \cos^2 \chi - t \sin^2 \chi = 0, \qquad (12.12)$$

since

$$\cos^2 \chi = \sin^2 \chi = \tfrac{1}{2}.$$

In the case when γ is large, however, the numerical values of the principal stresses t_1 and t_2 are no longer equal, the tensile stress (t_1)

in the direction of the major axis λ_1 increasing while the compressive stress $(-t_2)$ in the perpendicular direction diminishes with increasing strain (eqns (12.4)). This difference is accentuated by the decrease in χ with increasing strain, in accordance with eqn (12.3).

12.3. Torsion of a cylinder

The production of a state of homogeneous simple shear is not easily achieved experimentally, and direct measurements of the predicted normal stress components have not been carried out. There are, however, other types of strain which are more readily produced in which the effects of these normal stresses are directly manifested and may be quantitatively studied. The simplest of these is the torsion of a cylinder, in which the element of the body is in a state of simple shear.

To define the problem we consider a circular cylinder of radius a and height l which is twisted in such a way that the top surface is rotated through an angle $\theta = \psi l$ with respect to the lower surface, the height l remaining unchanged. (Fig. 12.4). This problem was

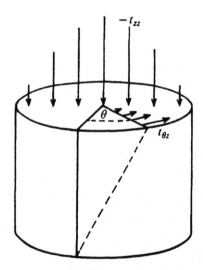

FIG. 12.4. Torsion of solid cylinder.

first treated by Rivlin (1948, 1949), who showed that the specified deformation can be maintained by forces applied to the end surfaces only. Since the state of strain is not homogeneous, but varies with the radial position r of the element considered, the corresponding distribution of the stress on the end surfaces is similarly non-uniform. To describe this stress distribution we make use of a cylindrical coordinate system (r, θ, z), in which the z-coordinate coincides with the axis of the cylinder. For simplicity we will assume the form of strain-energy function given by the statistical theory to apply in discussing the solution to this problem. The shear strain in an element at the radial position r being ψr, the corresponding shear stress is represented by the stress components $t_{\theta z}$ and $t_{z\theta}$, given by

$$t_{\theta z} = t_{z\theta} = G\psi r, \qquad (12.13)$$

where G is the shear modulus. The corresponding normal components of stress, t_{rr}, $t_{\theta\theta}$, and t_{zz} in the radial, circumferential, and axial directions, respectively, are obtainable from eqns (12.8), i.e.

$$t_{\theta\theta} - t_{rr} = G\psi^2 r^2, \qquad t_{zz} - t_{rr} = 0. \qquad (12.14)$$

If we consider a cylindrical shell of thickness dr, the circumferential stress $t_{\theta\theta}$ (Fig. 12.5) is equivalent to a hoop stress, which for

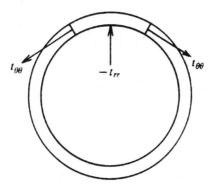

FIG. 12.5. Components of stress in cylindrical shell.

equilibrium must be balanced by an excess internal compressive stress. It follows that the radial component of (tensile) stress t_{rr} will be negative, and will increase in numerical value on proceeding towards the axis. The condition for equilibrium is represented by the differential equation

$$\frac{dt_{rr}}{dr} = \frac{t_{\theta\theta} - t_{rr}}{r} = G\psi^2 r. \qquad (12.15)$$

Integration of this equation, subject to the boundary condition that

$t_{rr} = 0$ when $r = a$, gives t_{rr}. Noting that $t_{rr} = t_{zz}$ (eqns(12.14)) the result becomes

$$t_{rr} = t_{zz} = \int_a^r G\psi^2 r \, dr = -\tfrac{1}{2}G\psi^2(a^2 - r^2). \qquad (12.16)$$

These components of stress therefore have a maximum value at the axis ($r = 0$) and decrease to zero at the outer surface of the cylinder ($r = a$).

The above analysis implies that the specified state of strain can be maintained by stresses applied to the end surfaces of the cylinder only, these stresses being distributed in accordance with eqns (12.13) and (12.16). It will be seen that the tangential stress $t_{\theta z}$ is proportional to the torsional strain ψ, and is identical in form to that given by the classical theory of elasticity for a material of shear modulus G. The normal stress t_{zz}, however, which varies as the *square* of the torsion, has no analogue on the classical theory.

The stress components represented by eqns (12.13) and (12.16), when integrated over the surface, correspond respectively to a couple M about the axis, and a normal tensile force N in the axial direction. Their values are as follows:

$$M = \int_0^a 2\pi r t_{\theta z} r \, dr = 2\pi G\psi \int_0^a r^3 \, dr$$

or

$$M = \tfrac{1}{2}\pi G\psi a^4, \qquad (12.17)$$

and

$$N = \int_0^a 2\pi r t_{zz} \, dr = -\pi G\psi^2 \int_0^a r(a^2 - r^2) \, dr$$

or

$$N = -\tfrac{1}{4}\pi G\psi^2 a^4. \qquad (12.18)$$

The normal force N, being negative, is equivalent to an axial compressive force or thrust. If this thrust is not applied, the cylinder will elongate in the axial direction.

12.4. Generalization of preceding results

The preceding treatment, based on the statistical theory, may readily be generalized to the case of a strain-energy function W of any desired form. The tangential and axial components of stress then become (Rivlin 1956)

$$t_{\theta z} = 2\psi r(\partial W/\partial I_1 + \partial W/\partial I_2),$$

$$t_{zz} = 2\psi^2\left(\int_a^r r\frac{\partial W}{\partial I_1}\,dr - r^2\frac{\partial W}{\partial I_2}\right),$$

(12.19)

leading to the following expression for the total couple and axial load,

$$M = 4\pi\psi\int_0^a r^3(\partial W/\partial I_1 + \partial W/\partial I_2)\,dr,$$

$$N = -2\pi\psi^2\int_0^a r^3(\partial W/\partial I_1 + 2\,\partial W/\partial I_2)\,dr.$$

(12.20)

In order to integrate these equations, it is necessary to know the form of W, since $\partial W/\partial I_1$ and $\partial W/\partial I_2$ will, in general, be functions of strain, and hence of r. For the particular case of the Mooney equation, for which $\partial W/\partial I_1$ and $\partial W/\partial I_2$ are both constants, the solution to (12.19) and (12.20) becomes

$$t_{\theta z} = 2\psi r(C_1 + C_2), \qquad M = \pi\psi a^4(C_1 + C_2),$$

(12.21)

and

$$t_{zz} = -\psi^2\{(C_1 - 2C_2)(a^2 - r^2) + 2a^2 C_2\},$$

(12.22a)

$$N = -\tfrac{1}{2}\pi\psi^2 a^4(C_1 + C_2).$$

(12.22b)

Eqns (12.21) correspond to the classical solution for a material of shear modulus $2(C_1 + C_2)$. Eqn (12.22a) represents a normal stress component which depends on the *square* of the torsional strain, and has no analogue on the classical theory. The form of distribution of this component is shown in Fig. 12.6(b). Since the axial and radial

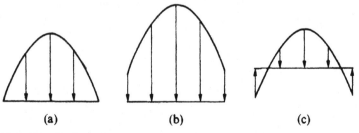

(a)   (b)   (c)

FIG. 12.6. Distribution of pressure over end of cylinder in torsion. (a) Statistical theory; (b) Mooney equation; (c) combined extension and torsion.

components of stress are not in this case equal (cf. eqns (12.11)), the vanishing of the radial stress component t_{rr} at the surface of the cylinder, required by the boundary conditions, does not require that the axial stress t_{zz} should also vanish at this point; in fact, for positive values of C_1 and C_2, it remains finite up to the edge ($r = a$). It is only for the special case $C_2 = 0$, corresponding to the statistical theory, that both components vanish simultaneously, and the distribution reduces to the simple parabolic form (Fig. 12.6(a)).

12.5. Experimental verification

Rivlin's experiment

Experiments confirming the existence and form of distribution of the axial stress in torsion were originally carried out by Rivlin (1947). A cylinder of vulcanized rubber was bonded on to metal end plates into which holes had been drilled at five different radial distances from the centre. On twisting the cylinder, while keeping the end plates at their initial distance apart, the rubber was observed to bulge out into the holes, thus indicating the development of an internal pressure. The height d of the bulge, which was assumed to be proportional to the pressure, was found to be proportional to the

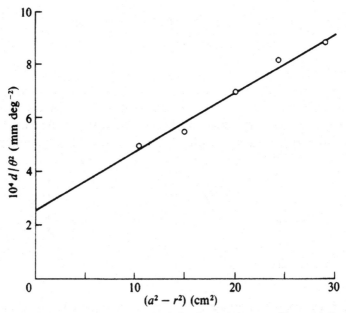

FIG. 12.7. Pressure distribution over end of cylinder in torsion. (From Rivlin 1947.)

square of the angle of rotation θ, for each value of r, in agreement with eqn (12.22a). Moreover, a plot of the ratio d/θ^2 against $a^2 - r^2$ yielded a straight line, as shown in Fig. 12.7. This result is consistent with eqn (12.22a), and from the values of the slope and intercept on the vertical axis the ratio C_2/C_1 was estimated to be equal to about $\frac{1}{7}$.

The experiments of Rivlin and Saunders

In a further more exact study of the torsion of a cylinder Rivlin and Saunders (1951) measured simultaneously both the total couple M and the total axial thrust $-N$ required to keep the axial length constant. The principle of their method is indicated in Fig. 12.8. The

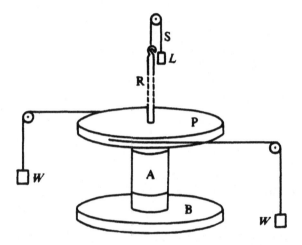

FIG. 12.8. Measurement of torsional couple and normal load on twisted cylinder. (Rivlin and Saunders 1951.)

rubber cylinder A, of length 1 in (2·54 cm) and diameter 1 in, was bonded to metal end pieces which were rigidly attached to the metal discs B and P. In carrying out the experiment the lower disc B was fixed in a horizontal plane and the weight of the upper disc counterbalanced by the load L connected to the supporting rod R via the cord S. Under these conditions the rubber specimen was unstrained. A known torque was then applied by means of weights W, causing a rotation of the upper disc and a slight increase in the axial length of the specimen. The latter was compensated by adjustment of the load L so as to restore the original length; the reduction in L was then equal to the normal thrust $-N$. Figs 12.9 and 12.10 show

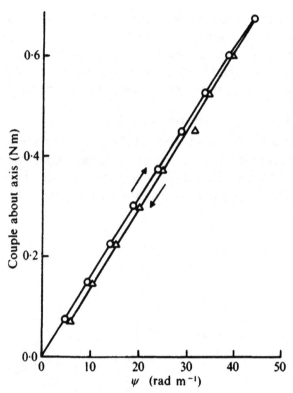

FIG. 12.9. Relation between torsional couple and amount of torsion, for cylinder of diameter 2·54 cm. (Rivlin and Saunders 1951.)

respectively the variation of torsional couple M with the amount of torsion ψ and the variation of $-N$ with ψ^2. These results confirm very beautifully the predictions of the theory, as represented by eqns (12.21) and (12.22b). The slopes of the lines in the figures give respectively $(C_1 + C_2)$ and $(C_1 + 2C_2)$; from the data shown the ratio C_2/C_1 was found to be $\frac{1}{8}$.

It should perhaps be noted that the interpretation of these data in terms of the Mooney equation is not inconsistent with the general criticisms of this equation advanced in Chapter 10, where it was argued that over a limited range of I_1 and I_2 the quantities $\partial W/\partial I_1$ and $\partial W/\partial I_2$ could be regarded as approximately constant. In the experiment of Rivlin and Saunders the maximum shear strain at the surface of the cylinder (ψr) was less than 0·6, corresponding to $\lambda_1 = 1·35$, giving $I_1 = I_2 = 3·37$, approximately. Over this small range of I_1 and I_2, i.e. from 3·0 to 3·37, consistency with the

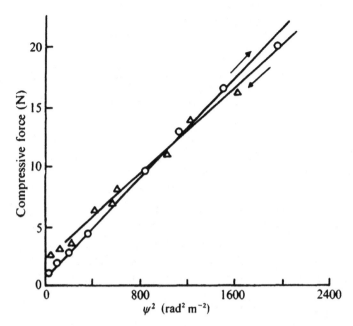

FIG. 12.10. Relation between normal load and square of torsion. (Rivlin and Saunders 1951.)

Mooney equation is to be expected. The ratio C_2/C_1 derived from the torsional data is also consistent with the ratio $(\partial W/\partial I_2)/(\partial W/\partial I_1)$ obtained from the pure homogeneous strain data of Fig. 10.8 (p. 224) in this range of strain.

12.6. Further problems in torsion

Combined torsion and extension

Rivlin (1949) has also considered the problem of torsion combined with axial extension (or compression) of a circular cylinder. This state of strain can also be maintained by forces applied to the end surfaces only. For the particular case of the Mooney strain-energy function the corresponding tangential and normal components of stress at the radial position r (in the unstrained state) are given by

$$t_{\theta z} = 2\psi\lambda^{\frac{3}{2}}r(C_1 + C_2)$$
$$t_{zz} = 2(\lambda^2 - 1/\lambda)(C_1 + C_2/\lambda) - \psi^2\{\lambda C_1(a^2 - r^2) + 2r^2 C_2\}, \quad (12.23)$$

where λ is is the axial extension ratio. The corresponding total couple and normal (tensile) load are

$$M = \pi\psi a^4(C_1 + C_2/\lambda)$$
$$N = 2\pi a^2(\lambda - 1/\lambda^2)(C_1 + C_2/\lambda) - \tfrac{1}{2}\pi\psi^2 a^4(C_1 + 2C_2/\lambda). \qquad (12.24)$$

Eqns (12.23) show the normal stress to be derived from two independent terms. The first term is a function of λ alone and is positive for $\lambda_1 > 1$. This corresponds to the ordinary tensile stress which, being independent of r, is uniform over the whole surface. The second term is proportional to the square of the torsion, and is negative (compressive); it corresponds to a parabolic distribution of pressure of the type represented in Fig. 12.6(b). The combination of these two terms may result in a reversal of the sign of the normal stress on proceeding from the axis to the edge of the cylinder (Fig. 12.6(c)). For sufficiently high values of λ the stress eventually becomes tensile at all points on the surface.

Torsion and extension of hollow cylinder

In the preceding torsional problems the specified state of strain could be maintained by forces applied to the end surfaces of the cylinder only. In the case of a hollow cylinder subjected to torsion, or to torsion plus axial extension, it is necessary in addition to apply forces to either the internal or external curved surfaces. These forces are necessitated by the radial stress component t_{rr}. Thus, for example, for the Mooney type of strain-energy function, a state of combined torsion and axial extension in a cylinder of external and internal radii a_1 and a_2 may be maintained by the application of a compressive stress to the inner surface of amount

$$-t_{rr} = \psi^2\lambda C_1(a_1^2 - a_2^2) \qquad (12.25)$$

with zero stress on the outer surface, together with stresses $t_{\theta z}$ and t_{zz} (Fig. 12.11) identical in form to the corresponding stresses in a solid cylinder of radius a_1, as given by eqns (12.21) and (12.22a). If the internal compressive stress is not applied, the cylinder will contract radially by an amount which, to a first approximation, is proportional to the square of the torsion.

Gent and Rivlin (1952) carried out experiments on tubes of vulcanized rubber subjected to combined extension and torsion. Means were provided for measuring the total couple, the internal pressure required to maintain the internal volume constant during

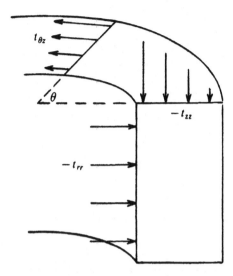

FIG. 12.11. Torsion of hollow cylinder.

the application of the torsion, and the change of internal volume which took place when no internal pressure was applied. They showed that the internal pressure required to maintain a constant volume was accurately proportional to the square of the torsion, as was also the change of volume in the absence of internal pressure.

12.7. Simultaneous extension, inflation, and shear of cylindrical annulus

Rivlin (1949) has also considered the successive application of the following strains to a tube of external and internal radii a_1 and a_2:

(1) a uniform simple extension parallel to the axis of the tube, in the ratio λ;

(2) a uniform inflation or radial expansion in which the length remains constant and the external and internal radii change to $\mu_1 a_1$ and $\mu_2 a_2$;

(3) a simple shear of the tube about its axis, in which each point rotates through an angle ϕ dependent only on its radial position;

(4) a simple shear of the tube in which each point moves parallel to the axis through a distance w which depends on its radial position.

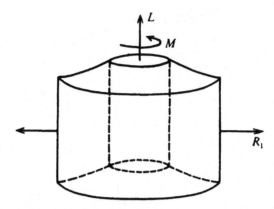

FIG. 12.12. Illustrating deformation of cylindrical annulus in combined torsion and shear.

The resultant form of the deformed tube is illustrated in Fig. 12.12. If the axis of the tube is in the z direction, the coordinates of a point initially at (r, θ, z) become $(\rho, \theta, \lambda z)$ by the transformations (1) and (2), $(\rho, \theta + \phi, \lambda z)$ by (3), and $(\rho, \theta + \phi, \lambda z + w)$ by (4), the values of ρ, ϕ and w being functions of r only. The precise form of the strain distribution, which is inhomogeneous, will depend on the form of the strain-energy function, and is worked out for the particular case of the Mooney form of strain-energy function. For this case the forces which have to be applied to the curved surfaces are:

(a)   normal tensile forces R_1 and R_2, respectively on the outer and inner surfaces, per unit unstrained length of axis;

(b)   an axial couple M (per unit unstrained axial length) acting on each of the curved surfaces;

(c)   a longitudinal tangential force L (per unit unstrained axial length) acting on each of the curved surfaces.

These forces are given by the following expressions:

$$R_1 = 2\pi a_1[(2/\lambda\mu_1)\{C_1 + (\mu_1^2 + \lambda^2\}C_2 + p_1\lambda\mu_1],$$
$$R_2 = 2\pi a_2[(2/\lambda\mu_2)\{C_1 + (\mu_2^2 + \lambda^2)\}C_2 + p_2\lambda\mu_2], \qquad (12.26)$$
$$M = 4\pi(C_1 + \lambda^2 C_2)K\phi_0'/\lambda,$$
$$L = 8\pi w_0'(C_1 + C_2/\lambda),$$

in which

$$K = a_1^2(\lambda\mu_1^2 - 1) = a_2^2(\lambda\mu_2^2 - 1)$$

and

$$\phi_0' = \phi_0/\ln(\mu_2/\mu_1), \qquad w_0' = w_0 \frac{(C_1 + \mu_2^2 C_2)a_2^2}{(C_1 + \mu_1^2 C_2)a_1^2},$$

ϕ_0 and w_0 being the values of ϕ and w at the outer boundary $r = a_2$. In addition, there are surface forces to be applied to the end surfaces of the tube which will not be considered here; these determine the values of p_1 and p_2.

12.8. Application of Ogden formulation

The alternative formulation of the strain-energy function proposed by Ogden (eqn (11.8)) may also be applied to problems involving torsion. This application has advantages in view of the restricted range of validity of the Mooney formula discussed in Chapter 10 and the difficulty of representing the actual properties of a rubber by any simple mathematical formula based on the use of the strain invariants I_1 and I_2. Applications to simple torsion of a cylinder and to combined torsion and axial extension have been worked out by Ogden and Chadwick (1972), who have also compared the stress distributions obtained on the basis of the three-term expression (11.25) with those derived from the Mooney equation.

Further problems treated on the basis of the Ogden formulation include the combined inflation and extension of a tube (Chadwick and Haddon 1972) and combined axial and torsional shear (Ogden, Chadwick, and Haddon 1973).

13

THERMODYNAMIC ANALYSIS OF GAUSSIAN NETWORK

13.1. Introduction

THE general significance of the thermodynamic study of the process of elastic deformation in rubberlike materials has already been discussed in Chapter 2, in which it was shown that from experiments on the temperature dependence of the stress in the extended state it is possible to derive the separate contributions due to the changes of internal energy and of entropy associated with the deformation process. The statistical theory in its elementary form attributes the elastic stress solely to the changes in configurational entropy of the molecular network; on this basis, therefore, the internal energy contribution to the stress should be zero. In practice, this expectation is not fully realized, there being in addition to the expected negative entropy contribution a significant contribution due to the internal energy changes. On the basis of the general thermodynamic treatment of Gee (1946a) and of Elliott and Lippmann (1945) it was further shown that the observed internal energy changes could plausibly be attributed to the changes of volume which accompany the application of the tensile stress to the rubber. As a corollary it was predicted that in an experiment carried out under constant volume conditions the internal energy contribution to the stress should be zero.

Two important developments, one experimental and the other theoretical, have necessitated a complete reappraisal of the situation. On the experimental side, Allen, Bianchi, and Price (1963) carried out the difficult task of actually measuring the internal energy changes accompanying an extension at constant volume, and showed that these were definitely not zero, but amounted to approximately 20 per cent of the total stress. On the theoretical side, Flory and his co-workers drew attention to the role of internal energy barriers to rotation about bonds *within* the single chain, and showed that these could lead to a significant contribution to the total free energy of deformation, even in the absence of volume changes

(Flory, Ciferri, and Hoeve 1960). Flory (1961) also developed a more rigorous theory of the Gaussian network, taking into account both volume changes and intramolecular internal energy effects, and thus provided a molecular or structural basis for the interpretation of the observed thermodynamic and thermoelastic effects under either constant volume or constant pressure conditions.

13.2. Force–Extension relation for Gaussian network

The case of simple extension, which is the most important from the practical standpoint, will first be considered. In the elementary theory the force–extension relation is of the form

$$f = NkT(\lambda - 1/\lambda^2), \tag{13.1}$$

where f is the force per unit unstrained area corresponding to the extension ratio λ referred to the unstrained length l_0 and N is the number of chains per unit volume. The derivation of this formula, as given in Chapter 4, makes use of the following assumptions:
   (1)   that the volume of the system is constant;
   (2)   that the mean-square chain vector length $\overline{r_i^2}$ in the unstrained state is the same as the mean-square vector length $\overline{r_0^2}$ of a corresponding set of free chains, the latter being given by $\overline{r_0^2} = \frac{3}{2}b^2$, where b^2 is the parameter in the Gaussian distribution formula (3.3) (p. 47)

The more accurate analysis of the Gaussian network takes account of the small volume changes which accompany the application of a stress. Under these conditions it becomes necessary to specify the extension ratio more precisely. The total deformation is made up of a uniform dilatation, corresponding to the change of volume, together with a distortional strain, or change of shape. It is convenient to define the extension ratio α with respect to the *undistorted state of volume V*, representing the volume in the final stressed state, i.e.

$$\alpha = l/l_i \tag{13.2}$$

where l is the length in the strained state and l_i the length in the undistorted state at the volume V. (It is important to note that l_i is *not* the unstrained length of the specimen.)

It is also necessary to replace N, the number of chains per unit volume, by ν/V, where ν is the total number of chains in the specimen, and V the volume.

These changes are purely geometrical in character, and are introduced as a direct consequence of the relaxation of the restriction implied by the first of the above two assumptions. The second of these assumptions, however, raises a more fundamental question. We have already seen from the discussion in Chapter 4 that this assumption lacks any firm theoretical basis. Of more importance for the present purpose, however, is the recognition that if the internal rotation about bonds within the molecule is not entirely free but is restricted by hindering potentials arising from steric interactions or other sources, the mean-square length of the free chains will itself be a function of temperature. It is therefore necessary to incorporate separately into the network theory the quantity $\overline{r_i^2}$, defined as the mean-square length of the chains in the undistorted state of the network at volume V, and the independent quantity $\overline{r_0^2}$, representing mean-square length of the corresponding set of free chains at a specified temperature T. The modified formula, which takes account of this distinction, as well as of the above considerations concerning the volume, is

$$\frac{f}{A_i} = \frac{vkT}{V} \frac{\overline{r_i^2}}{\overline{r_0^2}} \left(\alpha - \frac{1}{\alpha^2} \right), \qquad (13.3)$$

where f is the total force and A_i the undistorted cross-sectional area corresponding to the final volume V. Putting $A_i = V/l_i$, this may be written in the alternative form (Flory *et al.* 1960)

$$f = \frac{vkT}{l_i} \frac{\overline{r_i^2}}{\overline{r_0^2}} \left(\alpha - \frac{1}{\alpha^2} \right). \qquad (13.3a)$$

13.3. Stress–temperature relations

In order to derive the dependence of force on temperature under any specified conditions (e.g. constant volume or constant pressure), eqn (13.3a) must be expressed in terms of the observable quantity l rather than the extension ratio α. Introducing eqn (13.2) we therefore write

$$f = \frac{vkT}{l_i} \frac{\overline{r_i^2}}{\overline{r_0^2}} \left(\frac{l}{l_i} - \frac{l_i^2}{l^2} \right). \qquad (13.4)$$

In this expression $\overline{r_0^2}$ is a function of temperature but is independent of the volume of the specimen. On the other hand, $\overline{r_i^2}$ is a function of the volume of the specimen, but does not depend directly on temperature. Since the chain vector lengths for the cross-linked network in the undistorted state are proportional to the linear dimensions of the specimen it is clear that $\overline{r_i^2}$ is proportional to $V^{\frac{2}{3}}$. We have then, for a small change $d\overline{r_i^2}$ due to a change of volume dV,

$$d\overline{r_i^2}/\overline{r_i^2} = \tfrac{2}{3}(dV/V),$$

and hence

$$\frac{d(\ln \overline{r_i^2})}{dT} = \frac{1}{\overline{r_i^2}}\frac{d\overline{r_i^2}}{dT} = \frac{2}{3}\frac{1}{V}\frac{dV}{dT} = \frac{2}{3}\beta, \tag{13.5}$$

where β is the volume expansion coefficient of the unstrained rubber.

The above considerations enable the required force–temperature relations to be obtained. For a variation of temperature at constant length and constant *volume* the only quantities on the right-hand side of eqn (13.4) which vary are T and $\overline{r_0^2}$. Taking logarithms and differentiating we obtain

$$\left\{\frac{\partial \ln (f/T)}{\partial T}\right\}_{v,l} = -\frac{d \ln \overline{r_0^2}}{dT} \tag{13.6}$$

or

$$\left(\frac{\partial f}{\partial T}\right)_{v,l} = \frac{f}{T}\left[1 - T\frac{d \ln \overline{r_0^2}}{dT}\right]. \tag{13.6a}$$

In a similar manner, the differentiation of eqn (13.4) at constant length and constant *pressure*, i.e. under conditions permitting free expansion of volume, yields, in conjunction with eqn (13.5) and insertion of $\partial \ln l_i/\partial T = \beta/3$,

$$\left\{\frac{\partial \ln (f/T)}{\partial T}\right\}_{p,l} = -\frac{d \ln \overline{r_0^2}}{dT} - \frac{\beta}{\alpha^3 - 1} \tag{13.7}$$

or

$$\left(\frac{\partial f}{\partial T}\right)_{p,l} = \frac{f}{T}\left(1 - T\frac{d \ln \overline{r_0^2}}{dT} - \frac{\beta T}{\alpha^3 - 1}\right). \tag{13.7a}$$

From (13.6a) and (13.7a) the difference between the temperature coefficients of the force at constant volume and at constant pressure becomes

$$\left(\frac{\partial f}{\partial T}\right)_{v,l} - \left(\frac{\partial f}{\partial T}\right)_{p,l} = \frac{f\beta}{\alpha^3 - 1}. \tag{13.8}$$

13.4. Internal energy and entropy changes

Constant volume

The changes of internal energy and entropy associated with the deformation are obtainable from the above equations by the use of the general thermodynamic relations given in Chapter 2. Thus, for a constant-volume deformation,

$$\left(\frac{\partial S}{\partial l}\right)_{V,T} = -\left(\frac{\partial f}{\partial T}\right)_{V,l} = -\frac{f}{T}\left[1 - T\frac{\mathrm{d}\ln \overline{r_0^2}}{\mathrm{d}T}\right], \tag{13.9}$$

$$\left(\frac{\partial U}{\partial l}\right)_{V,T} = f + T\left(\frac{\partial S}{\partial l}\right)_{V,T} = fT\frac{\mathrm{d}\ln \overline{r_0^2}}{\mathrm{d}T}. \tag{13.10}$$

It is convenient to denote the internal energy contribution to the force, at constant volume, by the symbol f_e. The relative or fractional contribution due to internal energy then becomes

$$\frac{f_e}{f} = 1 - \frac{T}{f}\left(\frac{\partial f}{\partial T}\right)_{V,l} = T\frac{\mathrm{d}\ln \overline{r_0^2}}{\mathrm{d}T}. \tag{13.11}$$

The interpretation of this result is that the internal energy contribution to the force, in a deformation at constant volume, arises simply from the temperature dependence of the mean-square vector length of the free chains. Only in the case when the chain dimensions are independent of temperature, as is implicitly assumed in the elementary network theory, does this contribution vanish.

Conversely, the result provides a basis for the experimental determination of the temperature dependence of the mean-square chain dimensions.

Constant pressure

For a constant pressure experiment the corresponding expressions for the changes of entropy and heat content on deformation are

$$\left(\frac{\partial S}{\partial l}\right)_{p,T} = -\left(\frac{\partial f}{\partial T}\right)_{p,l} = -\frac{f}{T}\left(1 - T\frac{\mathrm{d}\ln \overline{r_0^2}}{\mathrm{d}T} - \frac{\beta T}{\alpha^3 - 1}\right), \qquad (13.12)$$

$$\left(\frac{\partial H}{\partial l}\right)_{p,T} = f + T\left(\frac{\partial S}{\partial l}\right)_{p,T} = fT\left(\frac{\mathrm{d}\ln \overline{r_0^2}}{\mathrm{d}T} + \frac{\beta}{\alpha^3 - 1}\right). \qquad (13.13)$$

Comparing eqn (13.13) with the corresponding eqn (13.10) for the internal energy change at constant volume, we obtain the difference

$$\left(\frac{\partial H}{\partial l}\right)_{p,T} - \left(\frac{\partial U}{\partial l}\right)_{V,T} = \frac{f\beta T}{\alpha^3 - 1}, \qquad (13.14)$$

which is directly related to the corresponding difference in the stress–temperature coefficients (eqn (13.8)). In the usual case when p is the atmospheric pressure, $\mathrm{d}H$ (which is equal to $\mathrm{d}U + p\,\mathrm{d}V$) is effectively identical with $\mathrm{d}U$. Eqn (13.14) thus gives the difference between the internal energy changes in experiments at constant pressure and at constant volume, respectively. This difference, which involves the expansion coefficient β, is due simply to the change of volume which accompanies the deformation under constant pressure conditions.

Eqn (13.14) provides a basis for the derivation of the internal energy changes at constant volume from experiments carried out at constant pressure. Combining eqns (13.8), (13.10), and (13.11) and putting $(\partial U/\partial l)_{V,T} = f_e$, we obtain

$$\frac{f_e}{f} = 1 - \frac{T}{f}\left(\frac{\partial f}{\partial T}\right)_{p,l} - \frac{\beta T}{\alpha^3 - 1}. \qquad (13.15)$$

Constant extension ratio

It is of interest to consider the temperature coefficient of tension at constant pressure and constant *extension ratio*. On the basis of general thermodynamics, Gee (1946a) derived the *approximate* relation

$$\left(\frac{\partial f}{\partial T}\right)_{p,\lambda} \simeq \left(\frac{\partial f}{\partial T}\right)_{V,l} \qquad (13.16)$$

(where λ is the extension ratio referred to the unstrained dimensions at the temperature T), from which eqn (2.14) (p. 35) immediately follows.

From eqn (13.3a) it is possible to obtain the more accurate relation

$$\left(\frac{\partial f}{\partial T}\right)_{p,\alpha} = \left(\frac{\partial f}{\partial T}\right)_{V,l} + \frac{f\beta}{3}. \tag{13.17}$$

Since $(\partial f/\partial T)_{p,\alpha}$ will differ from $(\partial f/\partial T)_{p,\lambda}$ to only a negligible extent, the additional term in (13.17) represents the error involved in Gee's approximate formula (13.16). The discrepancy arises from the fact that Gee assumed the rubber to be isotropically compressible under the action of a hydrostatic pressure in the strained state; this condition is represented by eqn (2.15) (p. 37), which may be written in the alternative form

$$3\left(\frac{\partial \ln l}{\partial \ln V}\right)_{T,f} = 1. \tag{13.18}$$

This relation, as he realized, is strictly valid only in the limit of zero strain ($\lambda - 1 \to 0$, $f \to 0$), in which case the two expressions (13.16) and (13.17) become equivalent. The quantitative evaluation of the anisotropy of compressibility can only be carried out on the basis of a specific structural model. The Gaussian network model, as represented by eqn (13.3a), may be shown to yield the result

$$3\left(\frac{\partial \ln l}{\partial \ln V}\right)_{T,f} = \frac{3}{\alpha^3 + 2}. \tag{13.19}$$

This formula was originally obtained by Khazanovitch (1959) and subsequently confirmed by Flory (1961).

13.5. Measurements at constant volume

With the advent of the theoretical developments discussed above, the direction and objectives of later experimental work became somewhat modified. It was seen that the most significant question from the theoretical standpoint was the evaluation of the temperature dependence of the mean-square chain dimensions, and that this was related directly to the internal energy contribution to the stress at constant volume through eqn (13.10). The latter quantity can be obtained either by direct measurements under constant volume conditions, or alternatively from measurements under constant pressure conditions, interpreted in the light of the theoretical relations (13.14) or (13.15). While most of the experimental work reported has been under the normal constant pressure conditions, particular interest is attached to the pioneer work of Allen and his associates, to which reference has already been made, on the direct measurement of internal energy changes under constant-volume conditions. Since these involve no theoretical assumptions, they are of the utmost value in establishing the validity of the theoretical relations between constant-volume and constant-pressure conditions arrived at on the basis of the Gaussian network theory.

Allen *et al.* (1963) measured the temperature dependence of the tension on a sample held at constant length whose volume was also held constant by the superposition of a hydrostatic pressure to counteract the normal expansion. Their apparatus is shown diagrammatically in Fig. 13.1. The rubber cylinder B, of length 6 in

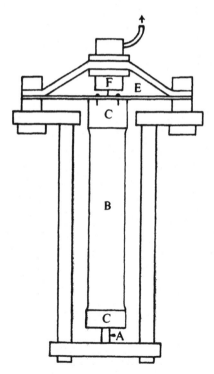

FIG. 13.1. Apparatus used for measurement of stress–temperature coefficient at constant volume (see text). (Allen, Bianchi, and Price, 1963.)

(15·24 cm), was bonded to stainless steel end pieces C attached at the top to a stiff spring E and at the bottom to a connecting rod A (whose length could be chosen to give any required extension), the whole being mounted on an Invar frame. By suitable choice of the length of the end pieces the residual expansion of the frame was practically eliminated, the resultant changes of length of the sample during the measurements being no more than 0·01 per cent. The stress was determined from the deflection of the spring, as measured by the transducer F, which was sensitive to a displacement of 0·015 μm. The whole assembly was contained in a steel pressure vessel capable of withstanding pressures up to 150 atm. The immersion liquid for the rubber was mercury, but for insulation purposes the steel strip and transducer were immersed in transformer oil floated on the mercury.

The value of the pressure required to maintain constancy of volume for the measurement of $(\partial f/\partial T)_{v,l}$ was obtained from subsidiary measurements of the 'thermal pressure coefficient' at constant length, $(\partial p/\partial T)_{v,l}$ at various values of the extension ratio.

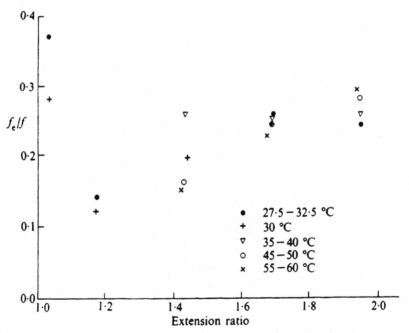

FIG. 13.2. Relative internal energy contribution to the tensile force (f_e/f) from measurements at constant volume. (Allen, Bianchi, and Price 1963.)

These experiments were carried out in a dilatometer. For completeness, measurements were also made of the bulk expansivity at constant length $V^{-1}(\partial V/\partial T)_{p,l}$. Neither of these coefficients showed any significant dependence on the extension ratio. The measurement of $(\partial f/\partial T)_{V,l}$ involved the variation of T in steps δT while the specimen was held at constant length, and the corresponding adjustment of p by an amount $\delta p = (\partial p/\partial T)_{V,l}\,\delta T$ at each step. The results obtained in this way for natural rubber are represented in Fig. 13.2 in terms of the relative internal energy contribution to the force at constant volume. Despite the considerable scatter, these results indicate that the internal energy contribution differs significantly from zero, the ratio f_e/f being of the order of 0·2.

The relation between the stress–temperature coefficients at constant volume and at constant pressure was also checked by independent measurements of $(\partial f/\partial T)_{p,l}$ and of the coefficients $(\partial f/\partial p)_{T,l}$ and $(\partial p/\partial T)_{V,l}$. These quantities are connected through the *exact* general thermodynamic relation

$$\left(\frac{\partial f}{\partial T}\right)_{V,l} = \left(\frac{\partial f}{\partial T}\right)_{p,l} + \left(\frac{\partial f}{\partial p}\right)_{T,l}\left(\frac{\partial p}{\partial T}\right)_{V,l}. \qquad (13.20)$$

The values calculated in this way are compared with the directly measured values in Table 13.1. In all cases the difference is within the experimental error.

TABLE 13.1

Direct and indirect determination of $(\partial U/\partial l)_{V,T}$ at 30 °C (Allen et al. 1963)

λ	f (kg cm^{-2})	$(\partial U/\partial l)_{V,T}$ calculated from eqn (13.20) (kg cm^{-2})	$(\partial U/\partial l)_{V,T}$ directly measured (kg cm^{-2})
1·690	3·56$_4$	0·86$_5$	0·91$_0$
1·43$_7$	2·65$_2$	0·52$_8$	0·52$_3$
1·18$_4$	1·42$_3$	0·16$_0$	0·20
1·03$_4$	0·31$_8$	0·09$_0$	0·11$_8$

Since eqn (13.20) is based on *exact* thermodynamic relations of a quite general character, this agreement has no bearing on the specific molecular model used in deriving eqn (13.14), which gives the explicit form of the relation between the internal-energy contributions to the stress at constant pressure and at constant volume, respectively. The measurements of Allen *et al.* (1963), however, may be used also to examine this relationship. The results are represented in Fig. 13.3, in terms of the difference between the

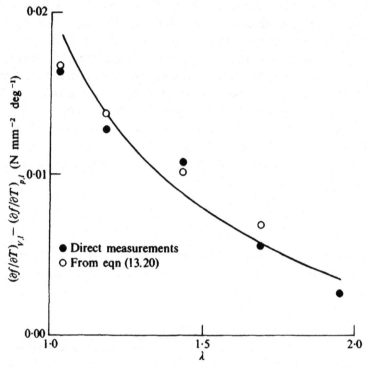

FIG. 13.3. Comparison of difference between $(\partial f/\partial T)_{V,l}$ and $(\partial f/\partial T)_{p,l}$ with theoretical formula (13.8), represented by continuous curve. (Allen, Bianchi, and Price 1963.)

corresponding stress–temperature coefficients. The continuous curve represents the theoretical relationship (13.8) and the experimental points were obtained either directly from the respective stress–temperature coefficients, or indirectly from the constant-pressure data together with the measured values of $(\partial f/\partial p)_{T,l}$ and $(\partial p/\partial T)_{V,l}$, in conjunction with eqn (13.20). In either case the predicted theoretical dependence on α is substantially verified.

Later work by Allen, Kirkham, Padget, and Price (1971), incorporating improvements in experimental accuracy has, however, suggested a possible slight discrepancy between those estimates of f_e/f based on the one hand on constant volume measurements or the equivalent exact relation (13.20), and on the other hand the values derived indirectly from constant pressure measurements through the use of eqn (13.15), the mean value obtained by the former method being 0.123 ± 0.022, compared with 0.18 ± 0.03 by the latter. This difference, though barely outside the experimental error, seems to be confirmed by later work referred to in § 13.13 and may possibly be connected with the deviations from the form of the force–extension curve predicted by the Gaussian network theory, discussed in Chapter 5. Against this, however, it is to be noted that these authors' directly measured values of f_e/f for various cross-linked rubbers in either the dry or swollen state showed no significant correlation with the deviations from the theoretical form of force–extension relation, as represented by the ratio C_2/C_1 of the constants in the Mooney equation.

13.6. Values of f_e/f

The quantity f_e/f, being related to the temperature dependence of the statistical length of the molecule through eqn (13.11), has a

TABLE 13.2

Values of f_e/f for natural rubber by different methods

Author	Method	f_e/f
Allen et al. (1971)	Constant V	0.12 ± 0.02
Allen et al. (1971)	Constant p	0.18 ± 0.03
Roe and Krigbaum (1962)	Constant p	$0.25–0.11$ depending on strain
Ciferri (1961)	Constant p	0.18
Barrie and Standen (1967)	From $(\partial l/\partial T)_{p,f}$	0.18
Shen (1969)	From $(\partial G/\partial T)_p$	0.15
Shen (1969)	From $(\partial l/\partial T)_{p,f}$	0.18 ± 0.02
Wolf and Allen (1974)	From $(\partial l/\partial T)_{p,f}$ Extension and compression	0.18 ± 0.2
Boyce and Treloar (1970)	Torsion (M_e/M)	0.126 ± 0.16
Gent and Kuan (1973)	Torsion (M_e/M)	0.17
Allen, Price, and Yoshimura (1975)	Calorimetry, extension	0.19 ± 0.02
Allen et al. (1975)	Calorimetry, torsion	$0.20_2 \pm 0.01_7$

characteristic value for any given polymer. Most of the data in the literature have been based on measurements of stress–temperature coefficients at constant pressure, using eqn (13.13), these being of course much more easily carried out than direct measurements at constant volume or their thermodynamic equivalent. A selection of data for a number of polymers is given in Table 13.2.

Possible dependence on strain

A number of authors have found the apparent values of f_e/f derived from constant-pressure data to be a function of the strain (Fig. 13.4). At very high extensions ($\lambda > 3$) the strong downward

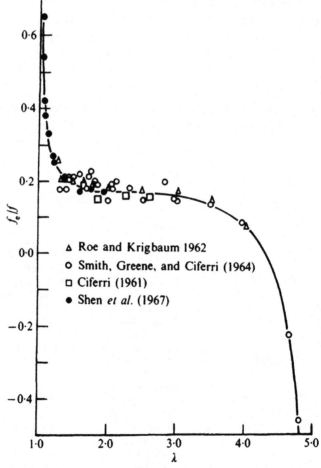

FIG. 13.4. Apparent dependence of f_e/f on strain. (From Shen *et al.* 1967.)

tendency, corresponding to a negative internal energy contribution to the stress, can reasonably be ascribed (in the case of natural rubber) to the incidence of strain-induced crystallization (cf. Fig. 2.9, p. 36). The sharp rise in f_e/f as the strain approaches zero, on the other hand, is not to be expected on theoretical grounds, and is almost certainly to be attributed to experimental difficulties, associated with the lack of perfect reversibility of the rubber (creep, etc.). This lack of reversibility makes an unambiguous definition of the unstrained length l_0 of the sample and hence of the strain parameter α in eqn (13.15) practically unattainable. Any error in l_0 has an exaggerated effect on the resulting value of f_e/f since the last term in eqn (13.15) tends to infinity as the *strain* $(\alpha - 1)$ tends to zero, whereas the preceding term tends to infinity as the *stress* tends to zero. If the zero of strain does not coincide exactly with the zero of stress errors of unlimited magnitude inevitably occur. The method is thus inherently unsatisfactory in the region of small strains.

13.7. Alternative experimental methods

Various indirect methods have been proposed in order to avoid the necessity for an accurate determination of the unstrained length inherent in the use of eqn (13.15) for the derivation of f_e/f from constant-pressure data. Shen (1969) recognized that the essential problem was the determination of the temperature coefficient of the *modulus*. He therefore plotted values of the force f, derived from force–temperature data at constant length, against $\lambda - 1/\lambda^2$, for different values of the temperature. Over the range of λ covered these plots were found to be linear, as illustrated in Fig. 13.5, indicating that the temperature dependence is not a function of the strain. The results were analysed on the basis of the equation

$$f = GA_0(\lambda - V/V_0\lambda^2) \tag{13.21}$$

in which G is equivalent to the shear modulus and is defined by

$$G = \frac{\nu k T}{V_0} \frac{\overline{r_i^2}}{\overline{r_0^2}}. \tag{13.22}$$

These equations are equivalent to (13.3) but are expressed in terms of the extension ratio λ referred to the stress-free state of volume

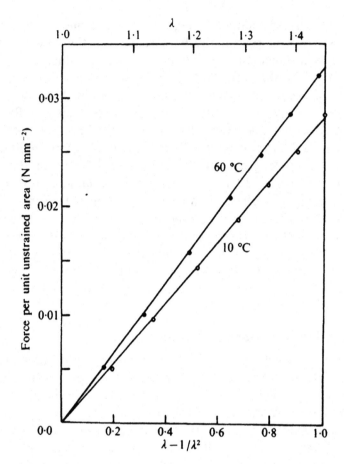

FIG. 13.5. Plots of f vs $\lambda - 1/\lambda^2$ at 60 °C and 10 °C for rubber swollen with 33·6 per cent hexadecane. (From Shen 1969.)

V_0 and area A_0. The equivalent expression for f_e/f then becomes

$$\frac{f_e}{f} = 1 - \frac{\mathrm{d}\ln G}{\mathrm{d}\ln T} - \frac{\beta T}{3}. \qquad (13.23)$$

The advantage of this method is that the last term in the above equation, namely, $-\beta T/3$, is not dependent on strain, in contrast to the corresponding term in (13.15).

A second alternative adopted by Shen involved the measurement of the change of *length* with temperature, at constant force, i.e.

$(\partial l/\partial T)_{p,f}$. For this Shen derives the relation[†]

$$\frac{1}{l}\left(\frac{\partial l}{\partial T}\right)_{p,f} = \frac{1}{l_0}\left(\frac{\partial l_0}{\partial T}\right)_{p,f} - \frac{\lambda^3-1}{\lambda^3+2}\left\{\frac{d\ln G}{dT} + \frac{2}{3l_0}\left(\frac{\partial l_0}{\partial T}\right)_{p,f}\right\}, \quad (13.24)$$

in which l_0 is the unstrained length at temperature T, so that $l_0^{-1}(\partial l_0/\partial T)_{p,f}$ is the linear expansivity in the unstrained state, and G is defined by eqn (13.22) above. Eqn (13.24) enables the temperature coefficient of shear modulus to be obtained, from which the value of f_e/f is then derived, using eqn (13.23).

According to eqn (13.24) a plot of $l^{-1}(\partial l/\partial T)_{p,f}$ against $(\lambda^3 - 1)/(\lambda^3+2)$ would be expected to be linear, the slope corresponding to $-d\ln G/dT$. The data for natural rubber in both dry and swollen states (Fig. 13.6) were found to be consistent with this relation, and

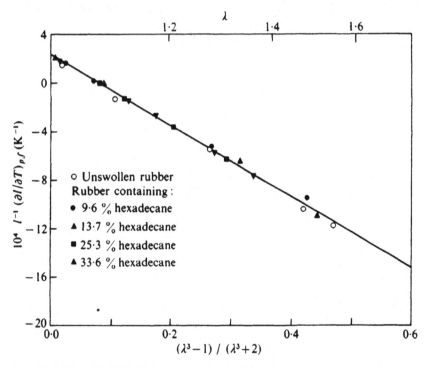

FIG. 13.6. Plot of $l^{-1}(\partial l/\partial T)_{p,f}$ against $(\lambda^3-1)/(\lambda^3+2)$ for unswollen rubber (0), and for rubber containing various percentages of hexadecane. (Shen 1969.)

[†] Shen uses the symbol β_L for $l^{-1}(\partial l/\partial T)_{p,f}$. In the present chapter this is liable to confusion with the volume expansivity at constant length, denoted by β_l. Similar considerations apply to the use of the symbol β_{lin} by Wolf and Allen (1975).

yielded a mean value of $0 \cdot 18$ for f_e/f, in close agreement with the mean value obtained from direct measurements of $d \ln G/dT$, namely $0 \cdot 15$. It is considered, however, that the former result is the more reliable.

Wolf and Allen (1975) have investigated the strain dependence of f_e/f, as derived from constant pressure measurements, for both uniaxial compression ($\lambda < 1$) and simple extension ($\lambda > 1$), using an apparatus which involved no discontinuity in the transition from compression to extension. They have also discussed the effect of deviations from the Gaussian theory and of experimental errors on the resultant values of f_e/f deduced from different forms of analysis. They show that the effect of deviations from the Gaussian theory involved in the application of eqn (13.15) is most serious at small strains but becomes unimportant for $\lambda > 1 \cdot 10$. The effect was shown to be most satisfactorily avoided as in Shen's experiments, by the measurement of the variation of length with temperature at constant force $l^{-1}(\partial l/\partial T)_{p,f}$. The value of f_e/f was then obtained from the relation

$$\frac{f_e}{f} = 1 + T\left[\frac{1}{l}\left(\frac{\partial l}{\partial T}\right)_{p,f} + \frac{1}{\lambda^3 - 1}\left\{\frac{3}{l}\left(\frac{\partial l}{\partial T}\right)_{p,f} - \beta_f\right\}\right], \quad (13.25)$$

in which β_f is the bulk expansion coefficient at constant force (which is not significantly different from β). The value of β_f is obtained from the measured linear expansivity at $f = 0$, thus avoiding any indeterminacy in the second term within the square brackets in eqn (13.25) as $\lambda \to 1$. For higher values of extension either of the equations

$$\frac{f_e}{f} = 1 - \frac{T}{f}\left(\frac{\partial f}{\partial T}\right)_{p,l} - \frac{\beta_l T}{\lambda^3 V/V_0 - 1}, \quad (13.26)$$

which is essentially identical to (13.15) or

$$\frac{f_e}{f} = 1 + T\left\{\left(\frac{\partial f}{\partial l}\right)_{T,p} \cdot \frac{1}{f}\left(\frac{\partial l}{\partial T}\right)_{p,l} - \frac{\beta_l}{\lambda^3 - 1}\right\} \quad (13.27)$$

were preferred. In these equations β_l is the bulk expansion coefficient at constant length (which again does not differ significantly from β).

The consistency in the values of f_e/f obtained by these different methods is shown in Fig. 13.7, which indicates that this quantity is

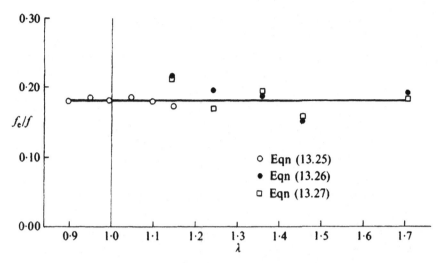

FIG. 13.7. Values of f_e/f in extension and uniaxial compression. (From Wolf and Allen 1975.)

independent of strain over the range of strain covered by the experiments. The mean value of f_e/f was 0.18 ± 0.02, in close agreement with the earlier work of Allen *et al.* (1971).

The methods of Shen and of Wolf and Allen are devices for eliminating the special significance of the absolute value of the strain either by utilizing functions which are not sensitive to the measurement of l_0, or by ensuring that the various equations containing terms involving l_0 are self-consistent. The results obtained for natural rubber by these various methods, together with others, are included in Table 13.2.

13.8. Theoretical analysis of torsion

Types of deformation which at first sight might be expected to circumvent the difficulties associated with volume changes are simple shear and the closely related torsion of a cylinder. As early as 1946 Meyer and van der Wyk subjected a sample of rubber contained in the annulus between concentric cylinders of metal to simple shear, and showed that the shear stress was approximately proportional to the absolute temperature, in accordance with the elementary statistical theory. However, as Flory *et al.* (1960) have pointed out, the simple interpretation of this result is misleading, and a more accurate analysis would require the inclusion of terms related to volume changes, which are normally neglected in the classical theory of elasticity.

The full analysis of the problem of torsion, first given by the author (Treloar 1969b), shows, however, that the effects of volume changes, though still significant, are quantitatively of a lower order of magnitude in this type of strain than in the case of simple extension (or compression). Experiments involving torsion therefore offer inherent advantages in the degree of accuracy attainable, compared with the more usual simple extension experiments. Apart from this, they provide a means of checking the important theoretical conclusion derived by Flory (1961) that the temperature coefficient of the stress, at *constant volume*, is directly related to the temperature dependence of the mean-square chain dimensions, *whatever the type of strain*—a conclusion which is illustrated in the case of simple extension by eqn (13.6a).

Let us consider a cylinder of unstrained radius a_0 subjected to combined axial extension and torsion about the axis of amount ψ given by

$$\psi = \phi/l, \tag{13.28}$$

where ϕ is the angle of rotation of the top surface with respect to the base of the cylinder, and l is the axial length in the strained state. This state of strain can be maintained by an axial load N together with a couple M about the axis. According to the accurate Gaussian network theory the latter is given by

$$M = \frac{\pi}{2} \frac{\nu k T}{V_u} \frac{\overline{r_i^2}}{\overline{r_0^2}} \psi a_0, \tag{13.29}$$

where ν is the total number of chains, V_u is the unstrained volume at temperature T and $\overline{r_0^2}$ and $\overline{r_i^2}$ have their previous significance (Treloar 1969b). This equation is derived on essentially the same basis as the corresponding equation (13.3) for simple extension, but incorporates the unstrained volume V_u rather than the strained or final volume V. It is to be noted that the value of the couple M (on the Gaussian theory) is independent of the axial extension.

For practical purposes the analysis may be restricted to the case when the length l is held constant while the temperature is varied. In this case the work done on the sample in a change of angular twist

$\mathrm{d}\phi$ is

$$\mathrm{d}W = M\,\mathrm{d}\phi - p\,\mathrm{d}V.$$

For a deformation at constant volume the couple may be expressed as the sum of internal energy and entropy components, i.e.

$$M = \left(\frac{\partial U}{\partial\phi}\right)_{V,l,T} - T\left(\frac{\partial S}{\partial\phi}\right)_{V,l,T}, \tag{13.30}$$

where, by analogy with the case of simple extension

$$\left(\frac{\partial S}{\partial\phi}\right)_{V,l,T} = -\left(\frac{\partial M}{\partial T}\right)_{V,l,\phi} \tag{13.31}$$

and

$$\left(\frac{\partial U}{\partial\phi}\right)_{V,l,T} = M + T\left(\frac{\partial S}{\partial\phi}\right)_{V,l,T} = M - T\left(\frac{\partial M}{\partial T}\right)_{V,l,\phi}. \tag{13.32}$$

For an experiment at constant pressure the corresponding relations are

$$\left(\frac{\partial S}{\partial\phi}\right)_{p,l,T} = -\left(\frac{\partial M}{\partial T}\right)_{p,l,\phi} \tag{13.33}$$

and

$$\left(\frac{\partial H}{\partial\phi}\right)_{p,l,T} = M - T\left(\frac{\partial M}{\partial T}\right)_{p,l,\phi}. \tag{13.34}$$

The values of $\partial M/\partial T$ corresponding to the Gaussian network theory are obtained by differentiation of eqn (13.29) under appropriate conditions. For constant l, constancy of ϕ is equivalent to constancy of ψ. Under constant volume conditions it follows that a_0 and r_i^2 are also constant; we therefore obtain

$$\left(\frac{\partial M}{\partial T}\right)_{V,l,\psi} = \left(\frac{\partial M}{\partial T}\right)_{V,l,\phi} = \frac{M}{T}\left(1 - T\frac{\mathrm{d}\ln\overline{r_0^2}}{\mathrm{d}T}\right). \tag{13.35}$$

Putting $(\partial U/\partial\phi)_{V,l,T} = M_e$, the relative internal energy contribution to the couple is then obtained from (13.32) in the form

$$\frac{M_e}{M} = T\frac{\mathrm{d}\ln\overline{r_0^2}}{\mathrm{d}T}, \tag{13.36}$$

which is precisely similar to the corresponding expression (13.11) for the case of simple extension, in accordance with Flory's general conclusion, referred to above.

For the case of constant pressure we note that $\overline{r_i^2}$ is proportional to $V^{\frac{2}{3}}$ while a_0^4 is proportional to $V^{\frac{4}{3}}$; the temperature dependence of M is then given by

$$\left(\frac{\partial M}{\partial T}\right)_{p,l,\psi} = \frac{M}{T}\left(1 + \beta T - T\frac{d\ln\overline{r_0^2}}{dT}\right), \tag{13.37}$$

where β is the volume expansion coefficient.

In order to obtain the value of M_e/M from an experiment conducted under constant pressure conditions eqns (13.36) and (13.37) may be combined to give

$$\frac{M_e}{M} = 1 - \frac{T}{M}\left(\frac{\partial M}{\partial T}\right)_{p,l,\psi} + \beta T. \tag{13.38}$$

Comparison with simple extension

Comparing the result represented by eqn (13.38) with the corresponding expression (13.15) for the case of simple extension, we see that all the terms in (13.38) remain finite as the torsional strain, and hence the couple M, approach zero. The evaluation of M_e/M is therefore not sensitive to the precise measurement of the unstrained dimensions, as is the evaluation of f_e/f from eqn (13.15). Thus, while a correction is still necessary to enable the internal energy change at constant volume to be derived from the stress–temperature coefficient measured under constant pressure conditions, this correction is much less critical, since it is independent of the magnitude of the strain.

From the physical standpoint this difference between the two types of strain may be understood very simply. In simple extension, under constant pressure conditions, the variation of temperature at constant length leads to a change in the strain, which in the case when the strain is small may be of the same order of magnitude as the strain itself. This therefore has a first-order effect on the stress–temperature coefficient, and is responsible, as we saw in Chapter 2, for the thermoelastic inversion phenomenon. In the case of torsion, on the other hand, a change of temperature has no affect on the torsional strain and affects the couple only through its affect or the radial dimension. The effect is therefore comparatively slight and is of the same relative magnitude for all values of the torsion. In particular, the sign of the stress–temperature coefficient is independent of strain, i.e. there is no thermoelastic inversion effect.

13.9. Experimental data for torsion

The above conclusions are fully borne out experimentally. Fig. 13.8 shows the relation between the torsional couple M and the temperature for a cylinder of rubber maintained at constant axial length and constant torsion ψ (Boyce and Treloar 1970). The quantity $(1/M)(\partial M/\partial T)_{p,l,\psi}$ derived from the slopes of these lines was found to be independent of the torsional strain, to within the accuracy of the experiment. It follows that the relative internal energy contribution to the stress is similarly independent of strain. The mean value of M_e/M obtained from these experiments was $0\cdot126\pm0\cdot016$, which may be compared with the value $0\cdot123\pm0\cdot022$ obtained by Allen *et al.* (1971) for f_e/f from constant-volume measurements.

Gent and Kuan (1973) have applied the torsional method to natural rubber, *trans*-isoprene and *cis*-butadiene polymers and high-density polyethylene in both the dry and swollen states. For the first three of these the values of M_e/M were unaffected by swelling, in agreement with earlier results. Polyethylene gave an anomalously high (negative) value, which though decreased by swelling, was still not in accord with the values given in Mark's review (Mark 1973). A more disturbing feature of their data, however, was the apparent reduction of M_e/M for the other three polymers with increasing *axial* extension. In the case of natural rubber, for example, the value of M_e/M fell linearly from about $0\cdot17$ at $\lambda=1$ to about $0\cdot07$ at $\lambda=3\cdot5$.

In considering these observations, it must be remembered that the equations used to calculate M_e/M are based on the Gaussian network theory. The authors consider that the observed strain dependence might be attributed to non-Gaussian (finite chain extensibility) effects. It could also possibly arise from stress-relaxation effects, in view of the fact that the measured couple is associated with the tangential stress component on a plane inclined at an angle less than 90° to the major stress axis and will therefore be sensitive to changes in the inclination of the stress ellipsoid which might occur as a result of stress relaxation.

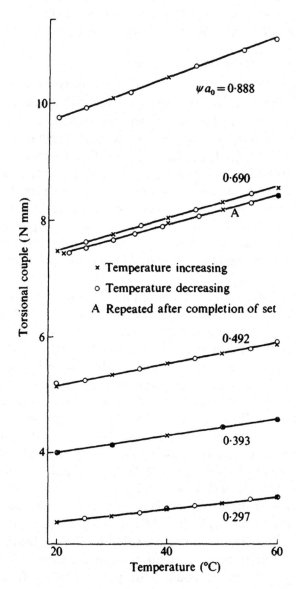

FIG. 13.8. Stress–temperature relations for torsion for various values of torsion parameter Ψa_0. (Boyce and Treloar 1970.)

The values for M_e/M reproduced in Table 13.2 refer to the smallest values of axial strain employed in these experiments.

13.10. Volume changes due to stress

Theoretical derivation

Since the original work of Gee (1946a) the question of the changes of volume which accompany the application of a stress, under constant pressure conditions, has been one of the central issues in the interpretation of the observed internal energy changes. It will be recalled (Chapter 2) that Gee attributed these internal energy changes *entirely* to the accompanying changes of volume, and endeavoured to confirm this relation by direct observations of volume changes. This approach, as we have seen, is inadequate and ignores the contribution to the total internal energy change arising from *intra*molecular forces, which is present even when the deformation takes place under the condition of constancy of volume. From this it follows immediately that the change of volume in a deformation at constant pressure should be related not to the total internal energy change, but to the difference between the internal energy changes at constant pressure and at constant volume, respectively, as given by eqn (13.14).

In the more refined treatment of the Gaussian network by Flory (1961) the total free energy is derived from two essentially independent terms, the first of which (A^*) is associated with the forces *between* the polymer molecules, considered as in the uncross-linked state, and the second (A_e) with the conformations of the cross-linked network. The first term is a function of volume and temperature only, and is independent of the network distortion; the second term is a function of the network deformation and of the temperature. The total stress on the system similarly consists of a component of hydrostatic pressure p^* arising from the term A^*, together with a three-dimensional stress system of the most general kind associated with the elastic deformation of the network. This model enables the volume change to be determined as a function of the strain. This is expressed by Flory, in the case of a simple extension, in terms of the 'dilation coefficient' (η) defined as

$$\eta = \left(\frac{\partial \ln V}{\partial \ln l}\right)_{T,p},$$ (13.39)

whose value is given by the expression

$$\eta = K_l \frac{\nu kT}{V} \left(\frac{V}{V_0}\right)^{\frac{2}{3}} \cdot \frac{1}{\alpha},$$ (13.40)

in which K_l is the bulk compressibility at constant length and V_0 is a 'reference' volume defined as that for which the equality $\overline{r_i^2} = \overline{r_0^2}$ would apply. The V_0, whose value is not determinable, may be eliminated by making use of eqn (13.4) to give

$$\eta = K_l \frac{\nu kT}{V} \frac{\overline{r_i^2}}{\overline{r_0^2}} \frac{1}{\alpha} = \frac{K_l fl}{V(\alpha^3 - 1)}.$$ (13.41)

The total volume change ΔV is obtained by integration from the unstrained length l_0 to the final length l. The result may be expressed in the alternative forms (Flory 1961; Christensen and Hoeve 1970):

$$\frac{\Delta V}{V} = K_l \frac{\nu kT}{V} \frac{\overline{r_i^2}}{\overline{r_0^2}} \left(1 - \frac{1}{\alpha}\right)$$ (13.42)

and

$$\frac{\Delta V}{V} = \frac{K_l fl}{V(1 + \alpha + \alpha^2)}.$$ (13.42a)

An alternative treatment of volume changes, based on exactly the same physical model, has been given by the author (Treloar 1969a). In this the equations are presented in terms of the actual unstrained volume V_u. This enables the pressure component p^* to be expressed directly in terms of the volume, and gives the principal stresses t_1, t_2, t_3 in a pure homogeneous strain in the form

$$t_i = \frac{V - V_u}{KV} + \frac{\nu kT}{V} \frac{\overline{r_i^2}}{\overline{r_0^2}} (\lambda_i^2 - 1) \qquad (i = 1, 2, 3).$$ (13.43)

For the case of simple extension eqn (13.43) reduces to a form which is substantially identical to (13.42) above.

13.11. Experimental examination

The above theoretical predictions have been tested both in the differential form (eqn (13.41)) by Allen and his associates, and in the integrated form (eqn (13.42a)) by Christensen and Hoeve. Allen *et al.* (1971), in the work already referred to, obtained the dilation coefficient from their measurements of $(\partial f/\partial p)_{l,T}$, making use of the thermodynamic relation

$$\left(\frac{\partial f}{\partial p}\right)_{l,T} = \left(\frac{\partial V}{\partial l}\right)_{p,T}. \tag{13.44}$$

For natural rubber vulcanizates the experimental values of dilation coefficient exceeded the theoretically predicted values by amounts which increased with increasing extension, the discrepancy being of the order of 100 per cent at the highest extension employed ($\alpha = 2\cdot0$). Only in the limit of zero strain ($\alpha = 1\cdot0$) was agreement with the theory obtained. Christensen and Hoeve (1971) obtained the total volume change directly by stretching the sample in a dilatometer by means of an electromagnetic device; in their case also the values were considerably in excess of the theory (Fig. 13.9).

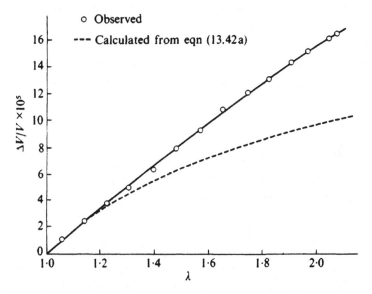

FIG. 13.9. Change of volume on extension. (Christensen and Hoeve 1970.)

Later measurements of dilation coefficient by Price and Allen (1973) yielded quantitatively similar deviations from the theory for both natural rubber and *cis*-polybutadiene rubber; for butyl rubber the data were inconclusive, since only relatively small values of extension ($\alpha = 1\cdot4$) were attained.

It must be concluded, therefore, that the Gaussian theory, in the form developed, does not in general provide a satisfactory quantitative basis for the interpretation of the volume changes on extension. As noted by Price and Allen, this conclusion is not altogether surprising, in view of the important deviations of the force–extension curve from the theoretical form discussed in detail in Chapter 11. Price and Allen have considered the possibility of modifying the theory by incorporating the Mooney form of force–extension relation; unfortunately, however, this means abandoning the Gaussian formula for the anisotropy of compressibility (eqn (13.19)) and reversion to the assumption of isotropic compressibility in the strained state (eqn (13.18)). The formula obtained in this way, namely,

$$\left(\frac{\partial \ln V}{\partial \alpha}\right)_{p,T} = K_f\left(\frac{C_1}{3}\frac{\alpha^3+2}{\alpha^2}+\frac{C_2}{\alpha^3}\right), \tag{13.45}$$

in which C_1 and C_2 are the Mooney constants and K_f is the volume compressibility at constant f, was found to give a satisfactory representation of the experimental data.

A simpler alternative formula derived on a purely empirical basis, which was found to be equally satisfactory, was the following:

$$\left(\frac{\partial \ln V}{\partial \alpha}\right)_{p,T} = K_l\left(\frac{C_1}{\alpha^2}+C_2\right). \tag{13.46}$$

13.12. Volume changes in torsion

In connection with the use of torsion as an alternative to simple extension for the experimental investigation of the internal energy component of the stress, it is of interest to examine theoretically the corresponding changes in volume in this type of strain. The treatment of this problem has been given by the author (Treloar 1969b). For a cylinder of unstrained radius a_0 subjected to an axial extension in the ratio β_3 together with a torsion ψ about the axis, referred to the length in the strained state, the expression for the relative volume change $\Delta V/V_u$ is found to be

$$\frac{\Delta V}{V_u} = K\frac{\nu kT}{V_u}\frac{\overline{r_i^2}}{r_0^2}\left\{\left(1-\frac{1}{\beta_3}\right)-\tfrac{1}{4}\beta_3\psi^2 a_0^2\right\}, \tag{13.47}$$

where K is the compressibility and V_u the volume in the unstrained state.

It is seen that this expression is made up of two independent terms of which the first (obtained by putting $\psi = 0$) represents the increase of volume due to the simple extension, and is equivalent to Flory's equation (13.41), while the second, which is necessarily negative, corresponding to a reduction of volume, represents the effect of the superimposed torsion. This second term, being proportional to the *square* of the torsion, has no analogue in the classical theory of elasticity.

The change of volume due to the torsion is thus seen to be a second-order effect. It may be regarded as having its origin in the normal components of stress which are present in simple shear or in torsion, and which are likewise proportional to the square of the corresponding strain (cf. Chapter 12). The total change of volume, in a combined extension and torsion, may be either positive or negative, depending on the relative magnitudes of the two components.

It is because the change of volume in torsion is of the second order, while that in extension is of the first order, that the differences of internal energy between a constant-pressure and a constant-volume deformation are also of a lower order in the case of torsion. It may in fact be proved explicitly, on the basis of general thermodynamics, that the calculated volume change leads exactly to a difference of internal energy of the form derived from the stress–temperature relations (13.35) and (13.37), respectively (Treloar 1969*b*). This analysis therefore confirms in detail the general deduction of Flory *et al.* (1960), referred to in § 13.8, concerning the inadequacy of the classical theory as a basis for thermoelastic analysis. It is remarkable, however, that even for infinitesimal strains, where the volume changes are negligible (in relation to the torsional strains), their relative contribution to the internal energy remains finite.

13.13. Calorimetric determination of internal-energy contribution to stress

The fundamental relation (for a system at constant pressure)

$$\Delta H = \Delta Q + \Delta W, \qquad (13.48)$$

where ΔQ is the heat absorbed by the system and ΔW the work performed by the external force, provides the basis for an alternative method of determining the internal energy change ΔH in the process of deformation by the direct measurement of ΔQ and ΔW. The method has not attracted much attention because of the difficulty of obtaining reliable calorimetric measurements on a system which may take several hours to attain mechanical equilibrium.

Recently, however, Allen, Price, and Yoshimura (1975) have studied both simple extension and torsion by this method, using for the purpose a Calvet micro-calorimeter, which measures the electrical energy input to a reference cell required to balance the heat output of the system being studied. After a preliminary high-temperature relaxation at the highest state of strain (tensile or torsional) the sample was cooled to the operating temperature (30 °C) and the deformation reduced by successive steps to zero, sufficient time being allowed at each stage for the system to reach thermal equilibrium.

In the torsional experiments the couple was substantially linear with respect to the torsion (ψ); the work W was therefore proportional to ψ^2. Correspondingly the heat absorbed was proportional to ψ^2 (Fig. 13.10). From these two sets of data the quantity

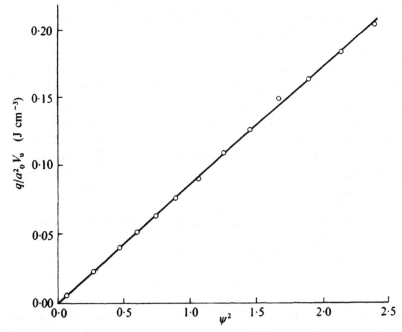

FIG. 13.10. Heat absorbed (q) on untwisting of cylinder from torsion ψ to unstrained state, plotted against ψ^2. (Allen *et al.* 1975.)

$(\partial H/\partial \phi)_{p,l,T}$, where $\phi = \psi l$, was directly obtained. From this the value of $(\partial U/\partial \phi)_{V,l,T}$ (or M_e) was then derived by the application of the equation (equivalent to (13.38))

$$\frac{M_e}{M} = \frac{1}{M}\left(\frac{\partial H}{\partial \phi}\right)_{p,l,T} + \beta T. \tag{13.49}$$

The mean value of M_e/M (for natural rubber) obtained in this way was $0.20_2 \pm 0.01_7$.

The simple extension data could not be so simply treated, since the relation between ΔH and α was not of any simple mathematical form, and the direct evaluation of f_e/f from the tangent at any point was rather inaccurate. The authors therefore used an indirect method which involved the dilation coefficient η, the details of which will not be reproduced here. The resulting values depended upon the formula employed to represent η, as shown in Table 13.3. The figures give some indication that the use of the Gaussian theory to convert constant-pressure data to constant-volume data may be subject to a slight error, but in view of the smallness of the differences obtained, and the somewhat insecure basis of eqns (13.45) and (13.46), a final judgement on this matter is hardly possible at present. The more important conclusion is that the direct calorimetric method provides an independent general confirmation of the results obtained from the more usual analysis of stress–temperature data.

TABLE 13.3

Formula for dilation coefficient	f_e/f
Gaussian (eqn (13.41))	0.19 ± 0.02
Eqn (13.45)	0.13 ± 0.02
Eqn (13.46)	0.14 ± 0.02

The calorimetric method has recently been applied by Price, Allen, and Yoshimura (1975) to cis-polybutadiene) rubber. For extension they obtained $f_e/f = 0.11$, while for torsion the corresponding value was 0.14, giving a mean of 0.12_5.

13.14. Temperature dependence of chain dimensions

From the experimentally determined value of the relative internal energy contribution to the stress f_e/f, at constant volume, the temperature dependence of the statistical chain length r_0 for the free molecule follows immediately from eqn (13.11). Numerical values for this quantity for selected polymers are given in the accompanying table (for a more comprehensive list see Mark (1973)).

It will be seen that for natural rubber the temperature coefficient of $\overline{r_0^2}$ is positive, which implies that energetic interactions within the chain act so as to reduce the mean-square length; i.e. a shorter length has a lower internal energy. For polythene, on the other hand, the temperature coefficient is negative, indicating that the extended (*trans*-) conformation of the chain is energetically favoured. Butyl rubber (polyisobutylene) also shows a small affect in the same sense.

TABLE 13.4

Values of f_e/f and $\mathrm{d}\ln \overline{r_0^2}/\mathrm{d}T$ for various polymers

Polymer	f_e/f	$10^3 \, \mathrm{d}\ln \overline{r_0^2}/\mathrm{d}T$ (K^{-1})	Reference
Natural rubber	0·12	0·38	Allen *et al.* (1971)
trans-polyisoprene	0·17	0·53	Barrie and Standen (1967)
Butyl rubber	−0·08	−0·26	Allen *et al.* (1968)
Silicone rubber	0·25	0·82	Allen *et al.* (1969)
cis-polybutadiene	0·10	0·31	Shen, Chen, Cirlin, and Gebhard (1971)
cis–polybutadiene	0·12$_5$	0·41	Price *et al.* (1975)
Polyethylene	−0·42	−0·97	Ciferri, Hoeve, and Flory (1961)

It is of interest to compare the thermoelastic estimates of the temperature dependence of $\overline{r_0^2}$ with estimates derived from solution viscosities. Such a comparison has been made, in the case of polyethylene, by Ciferri, Hoeve, and Flory (1961). For solvents which have no specific effect on the statistical form of the polymer molecule (the so-called 'θ-solvents') values of $10^3 \, \mathrm{d}\ln \overline{r_0^2}/\mathrm{d}T$ in the range $(-1\cdot1\pm0\cdot5)\mathrm{K}^{-1}$ have been obtained in this way. This figure may be compared with the same authors' thermoelastic data, which gave $-0\cdot97 \, \mathrm{K}^{-1}$ for polythene in the unswollen state and $-1\cdot01 \, \mathrm{K}^{-1}$ and $-1\cdot16 \, \mathrm{K}^{-1}$ for the swollen state, using two different solvents (see Table 13.4).

13.15. Conclusion

From the material presented in this chapter it will be seen that the conclusions derived from the application of thermodynamics to the analysis of the elastic deformation of rubber are wide-ranging in their implications, and spread far beyond the rather limited objectives of earlier studies in this field.

The most important conclusion is that the deformation of the chain is not purely entropic, but also involves significant changes in the intramolecular energy. It follows that any detailed quantitative treatment of the statistical properties of real molecules must take account of the effect of energetic interactions between neighbouring groups along the chain. This is in itself a large subject, which cannot be elaborated here; the interested reader is referred to the monograph by Flory (1969). In the case of a network, this intramolecular internal energy is revealed in its simplest form when the deformation is carried out at constant volume. In the more usual constant-pressure type of deformation there will be an additional contribution to the internal energy arising from the small but significant change in volume of the system.

The more recent developments of the Gaussian network theory, particularly by Flory, have enabled the change in volume to be predicted, and have provided a more accurate basis for the interpretation of constant-pressure data. While there is evidence that the predictions of the theory, particularly with respect to the changes of volume due to the stress, are not wholly in accord with experiment, there is in general a fair degree of consistency between thermoelastic results obtained under different conditions (constant pressure and constant volume) and by the use of different types of strain (extension, compression, torsion, etc.), when these are interpreted on the basis of the theory. Finally, the independent estimation of internal energy changes from the direct calorimetric measurement of the heat of deformation provides a very significant check on the basic soundness of the rather elaborate thermodynamic reasoning involved in the thermoelastic studies.

REFERENCES

ALEKSANDROV, A. P. and LAZURKIN, Y. S. (1939). *J. tech. Phys. USSR* **9**, 1249.

ALEXANDER, H. (1968). *Int. J. engng Sci.* **6**, 549.

ALLEN, G., BIANCHI, U., and PRICE, C. (1963). *Trans. Faraday Soc.* **59**, 2493.

——, KIRKHAM, M. J., PADGET, J., and PRICE, C. (1971). *Trans. Faraday Soc.* **67**, 1278.

——, PRICE, C., and YOSHIMURA, N. (1975). *J. Chem. Soc. Faraday Trans.* I, **71**, 548.

ANDREWS, E. H., OWEN, P. J., and SINGH, A. (1971). *Proc. R. Soc.* A**324**, 79.

ANTHONY, R. L., CASTON, R. H., and GUTH, E. (1942). *J. phys. Chem.* **46**, 826.

BARRIE, J. A. and STANDEN, J. (1967). *Polymer* **8**, 97.

BEKKEDAHL, N. (1934). *Natn. Bur. Stand. J. Res., Wash.* **13**, 410.

BIDERMAN, V. L. (1958). *Rascheti na prochnost*, Moscow.

BLANKS, R. F. and PRAUSNITZ, J. M. (1964). *Ind. eng. Chem. Funda.* **3**, 1.

BLOOMFIELD, G. F. (1946). *J. Polymer Sci.* **1**, 312.

BOBEAR, W. J. (1966). *J. Polymer Sci.* A2, **4**, 299.

BOYCE, P. H. and TRELOAR, L. R. G. (1970). *Polymer* **11**, 21.

BUNN, C. W. (1942). *Proc. R. Soc.* A**180**, 40.

——, and DAUBENY, R. DE P. (1954). *Trans. Faraday Soc.* **50**, 1173.

CASE, L. C. (1960). *J. Polymer Sci.* **45**, 397.

CHADWICK, P. and HADDON, E. W. (1972). *J. Inst. Maths. Appl.* **10**, 258.

CHRISTENSEN, R. G. and HOEVE, C. A. J. (1970). *J. Polymer Sci.* A1, **8**, 1503.

CIFERRI, A. (1961). *Makro. Chemie* **43**, 152.

—— and FLORY, P. J. (1959). *J. appl. Phys.* **30**, 1498.

——, HOEVE, C. A. J., and FLORY, P. J. (1961). *J. Am. chem. Soc.* **83**, 1015.

DART, S. L., ANTHONY, R. L., and GUTH, E. (1942). *Ind. engng Chem.* **34**, 1340.

DENBIGH, K. G. (1940). *Trans. Faraday Soc.* **36**, 936.

DI MARZIO, E. A. (1962). *J. chem. Phys.* **36**, 1563.

ELLIOTT, D. A. and LIPPMANN, S. A. (1945). *J. appl. Phys.* **16**, 50.

FARADAY, M. (1826). *Q.J.Sci.* **21**, 19.

FARMER, E. H. and SHIPLEY, F. W. (1946). *J. Polymer Sci.* **1**, 293.

FIKENTSCHER, H. and MARK, H. (1930). *Kautschuk* **6**, 2.

FLORY, P. J. (1942). *J. chem. Phys.* **10**, 51.

—— (1944). *Chem. Rev.* **35**, 51.

—— (1946). *Ind. engng Chem.* **38**, 417.

—— (1947). *J. chem. Phys.* **15**, 397.

—— (1950). *J. chem. Phys.* **18**, 108.

—— (1953). *Principles of polymer chemistry.* Cornell, Ithaca, N.Y.
—— (1961). *Trans. Faraday Soc.* **57**, 829.
—— (1969). *Statistical mechanics of chain molecules.* Interscience, New York.
——, CIFERRI, A., and HOEVE, C. A. J. (1960). *J. Polymer Sci.* **45**, 235.
——, RABJOHN, N., and SHAFFER, M. C. (1949). *J. Polymer Sci.* **4**, 225.
—— and REHNER, J. (1943). *J. chem. Phys.* **11**, 512.
——, —— (1944). *J. chem. Phys.* **12**, 412.
GEE, G. (1942). *Trans. Faraday Soc.* **38**, 418.
—— (1943). *Inst. Rubb. Ind. Trans.* **18**, 266.
—— (1946a). *Trans. Faraday Soc.* **42**, 585.
—— (1946b). *Trans. Faraday Soc.* **42B**, 33.
—— (1947). *J. Polymer Sci.* **2**, 451.
—— and ORR, W. J. C. (1946). *Trans. Faraday Soc.* **42**, 507.
—— and TRELOAR, L. R. G. (1942). *Trans. Faraday Soc.* **38**, 147.
GEIL, P. H. (1963). *Polymer single crystals.* Interscience, New York.
GENT, A. N. (1969). *Macromolecules* **2**, 262.
—— and KUAN, T. H. (1973). *J. Polymer Sci. (Polymer Phys.)* **11**, 1723.
—— and RIVLIN, R. S. (1952). *Proc. phys. Soc.* **B65**, 118, 487, and 645.
—— and THOMAS, A. G. (1958). *J. Polymer Sci.* **28**, 625.
—— and VICKROY, V. V. (1967). *J. Polymer Sci.* **A2**, **5**, 47.
GOPPEL, J. M. (1949). *Appl. Sci.* **A1**, 3 and 18.
GORDON, M., KUCHARIK, S., and WARD, T. C. (1970). *Collect. Czech. chem. Commun.* **35**, 3252.
GOUGH, J. (1805). *Mem. lit. phil. Soc. Manchester* **1**, 288.
GUGGENHEIM, E. A. (1952). *Mixtures.* Oxford University Press.
GUMBRELL, S. M., MULLINS, L., and RIVLIN, R. S. (1953). *Trans. Faraday Soc.* **49**, 1495.
GUTH, E., JAMES, H. M., and MARK, H. (1946). *Advances in colloid science II*, p. 253. Interscience, New York.
—— and MARK, H. (1934). *Monatsh. Chem.* **65**, 93.
HALL, P. (1927). *Biometrika* **19**, 240.
HALLER, W. (1931). *Kolloidzschr.* **56**, 257.
HART-SMITH, L. J. (1966). *J. appl. Maths. Phys.* **17**, 608.
HILDEBRAND, J. H. and SCOTT, R. L. (1950). *The solubility of non-electrolytes.* Reinhold, New York.
HILL, J. L. and STEPTO, R. F. T. (1971). *Trans. Faraday Soc.* **67**, 3202.
HOCK, L. (1925). *Z. Elektrochem.* **31**, 404.
HOUWINK, R. (1949). *Elastomers and plastomers*, p. 339. Elsevier, Amsterdam.
HUGGINS, M. L. (1942). *Ann. N. Y. Acad. Sci.* **43**, 1.
IRWIN, J. O. (1927). *Biometrika* **19**, 225.
ISIHARA, A., HASHITSUME, N., and TATIBANA, M. (1951). *J. chem. Phys.* **19**, 1508.
JAMES, H. M. and GUTH, E. (1943). *J. chem. Phys.* **11**, 455.
——, —— (1947). *J. chem. Phys.* **15**, 669.
JENKINS, F. A. and WHITE, H. E. (1957). *Fundamentals of optics.* McGraw-Hill, New York.

JONES, D. F. and TRELOAR, L. R. G. (1975). *J. Phys. D. (Appl. Phys.)* **8**, 1285.

JOULE, J. P. (1859). *Phil. Trans. R. Soc.* **149**, 91.

KARRER, E. (1933). *Protoplasma* **18**, 475.

KATZ, J. R. (1925). *Chem. Z.* **49**, 353.

—— (1933). *Trans. Faraday Soc.* **29**, 297.

KAWABATA, S. (1973). *J. macromol. Sci.* B8, 3–4, 605.

KELVIN, LORD (THOMSON, W.) (1857). *Q.J. Pure appl. Math.* **1**, 57.

KHAZANOVITCH, T. N. (1959), *J. appl. Phys.* **30**, 948.

KRAUS, G. and MOCZVGEMBA, G. A. (1964). *J. Polymer Sci.* A2, 277.

KRIGBAUM, W. R. and KONEKO, M. (1962). *J. chem. Phys.* **36**, 99.

—— and ROE, R. J. (1965). *Rubber Chem. Technol.* **38**, 1039.

KUHN, W. (1934). *Kolloidzschr.* **68**, 2.

—— (1936). *Kolloidzschr.* **76**, 258.

—— and GRÜN, F. (1942). *Kolloidzschr.* **101**, 248.

—— and KUHN, H. (1946). *Helv. chim. Acta* **29**, 1095.

LE BRAS, J. (1965). *Introduction to rubber*, p. 8. Maclaren, London.

LOKE, K. M., DICKINSON, M., and TRELOAR, L. R. G. (1972). *Polymer* **13**, 203.

LOVE, A. E. H. (1934). *The mathematical theory of elasticity*. Cambridge University Press.

MACK, E. (1934). *J. Am. chem. Soc.* **56**, 2757.

MARK, J. E. (1973). *Rubber Chem. Technol.* **46**, 593.

MEISSNER, B. (1967). *J. Polymer Sci.* C16, 781 and 793.

MEMMLER, K. (1934). *The Science of rubber*. Reinhold, New York.

MEYER, K. H. (1939). *Helv. chim. Acta* **22**, 1362.

—— and FERRI, C. (1935). *Helv. chim. Acta* **18**, 570.

—— and VAN DER WYK, A. J. A. (1946). *Helv. chim. Acta* **29**, 1842.

——, VON SUSICH, G., and VALKO, E. (1932). *Kolloidzschr.* **59**, 208.

MILLS, N. J. and SAUNDERS, D. W. (1968). *J. Macromol. Sci.* B2, 369.

MOONEY, M. (1940). *J. appl. Phys.* **11**, 582.

MOORE, C. G. and WATSON, W. F. (1956). *J. Polymer Sci.* **19**, 237.

MORGAN, R. J. and TRELOAR, L. R. G. (1972). *J. Polymer Sci.* A2, **10**, 51.

MORRELL, S. H. and STERN, J. (1953). *Inst. Rubb. Ind. Trans.* **28**, 269.

MORRIS, M. C. (1964). *J. appl. Polymer Sci.* **8**, 545.

MULLINS, L. (1947). *Inst. Rubb. Ind. Trans.* **22**, 235.

—— (1956). *J. Polymer Sci.* **19**, 225.

—— (1959*a*). *J. appl. Polymer Sci.* **2**, 1.

—— (1959*b*). *J. appl. Polymer Sci.* **2**, 257.

OBATA, Y., KAWABATA, S. and KAWAI, H. (1970). *J. Polymer Sci.* A2, **8**, 903.

OGDEN, R. W. (1972). *Proc. R. Soc.* A326, 565.

—— and CHADWICK, P. (1972). *J. Mech. Phys. Solids* **20**, 77.

——, ——, and HADDON, E. W. (1973). *Q.J. Mech. appl. Math.* **26**, 23.

OSTWALD, W. (1926). *Kolloidzschr.* **40**, 58.

PORTER, M. (1967). *J. appl. Polymer Sci.* **11**, 2255.

PRAGER, S. and FRISCH, H. L. (1967). *J. chem. Phys.* **46**, 1475.

PRICE, C. and ALLEN, G. (1973). *Polymer* **14**, 576.

———, ———, DE CANDIA, F., KIRKHAM, M. C., and SUBRAMANIAM, A. (1970). *Polymer* **11**, 486.

———, ———, and YOSHIMURA, N. (1975). *Polymer* **16**, 261.

RIVLIN, R. S. (1947). *J. appl. Phys.* **18**, 444.

——— (1948). *Phil. Trans. R. Soc.* A**241**, 379.

——— (1956). *Rheology, theory and applications* (Ed. F. R. Eirich), p. 351. Academic Books, London.

——— and SAUNDERS, D. W. (1951). *Phil. Trans. R. Soc.* A**243**, 251.

ROE, R. J. and KRIGBAUM, W. R. (1962). *J. Polymer Sci.* **61**, 167.

———, ——— (1963). *J. Polymer Sci.* A**1**, 2049.

ROTH, F. L. and WOOD, L. A. (1944). *J. appl. Phys.* **15**, 749.

SACK, R. A. (1956). *J. chem. Phys.* **25**, 1087.

ST. PIERRE, L. E., DEWHURST, H. A., and BUECHE, A. M. (1959). *J. Polymer Sci.* **36**, 105.

SAUNDERS, D. W. (1950). *Nature, Lond.* **165**, 360.

——— (1956). *Trans. Faraday Soc.* **52**, 1414 and 1425.

——— (1957). *Trans. Faraday Soc.* **53**, 860.

———, LIGHTFOOT, D. R., and PARSONS, D. A. (1968). *J. Polymer Sci.* A**2**, **6**, 1183.

SCANLAN, J. (1960). *J. Polymer Sci.* **43**, 501.

SCHWARZ, J. (1973). *Kolloidzschr. u.Z. Polymere* **251**, 215.

SHEEHAN, C. J. and BISIO, A. L. (1966). *Rubber Chem. Technol.* **39**, 149.

SHEN, M., (1969). *Macromolecules* **2**, 358.

———, CHEN, T. Y., CIRLIN, E. H., and GEBBARD, H. M. (1971). *Polymer networks, structure and mechanical properties* (eds. A. J. Chomff and S. Newman). Plenum Press, New York.

———, MCQUARRIE, D. A., and JACKSON, J. L. (1967). *J. appl. Phys.* **38**, 791.

SHVARTS, A. G. (1958). *Rubber Chem. Technol.* **31**, 691.

SMITH, K. J., GREENE, A., and CIFERRI, A. (1964). *Kolloidzschr.* **194**, 49.

——— and PUETT, D. (1966). *J. appl. Phys.* **37**, 346.

THIBODEAU, W. E. and MCPHERSON, A. T. (1934). *Natn. Bur. Stand. J. Res., Wash.* **13**, 887.

THOMAS, A. G. (1955). *Trans. Faraday Soc.* **51**, 569.

TOMPA, H. (1956). *Polymer solutions*, Butterworth, London.

TRELOAR, L. R. G. (1941). *Trans. Faraday Soc.* **37**, 84.

——— (1943*a*). *Trans. Faraday Soc.* **39**, 36.

——— (1943*b*). *Trans. Faraday Soc.* **39**, 241.

——— (1943*c*). *Proc. phys. Soc.* **55**, 345.

——— (1944*a*). *Trans. Faraday Soc.* **40**, 59.

——— (1944*b*). *Trans. Faraday Soc.* **40**, 109.

——— (1946*a*). *Trans. Faraday Soc.* **42**, 77.

——— (1947). *Trans. Faraday Soc.* **43**, 277 and 284.

——— (1948). *Proc. phys. Soc.* **60**, 135.

——— (1950*a*). *Proc. Roy. Soc.* A**200**, 76.

——— (1950*b*). *Trans. Faraday Soc.* **46**, 783.

—— (1953). *Trans. Faraday Soc.* **49**, 816.
—— (1954). *Trans. Faraday Soc.* **50**, 881.
—— (1969*a*). *Polymer* **10**, 279.
—— (1969*b*). *Polymer* **10**, 291.
—— (1972). *Polymer* **13**, 195.
TSCHOEGL, N. W. (1971). *J. Polymer Sci.* A1, **9**, 1959.
VALANIS, K. C. and LANDEL, R. F. (1967). *J. appl. Phys.* **38**, 2997.
VOLKENSTEIN, M. V. (1963). *Configurational statistics of polymer chains.* Interscience, New York.
WALL, F. T. (1942). *J. chem. Phys.* **10**, 485.
—— (1943). *J. chem. Phys.* **11**, 67.
—— and FLORY, P. J. (1951). *J. chem. Phys.* **19**, 1435.
WANG, M. C. and GUTH, E. (1952). *J. chem. Phys.* **20**, 1144.
WÖHLISCH, E. (1926). *Verh. Phys.-Med. Ges. Wurz.* N.F. **51**, 53.
WOLF, F. P. and ALLEN, G. (1975). *Polymer* **16**, 209.
WOOD, L. A. (1946). *Natn. Bur. Stand. J. Res. Wash.* **36**, 489.
—— and ROTH, F. L. (1944). *J. appl. Phys.* **15**, 781.

AUTHOR INDEX

SUBJECT INDEX

bond polarizabilities, 203

chain dimensions,
statistical theory, 42 ff.; Gaussian distribution, 46; polyisoprene chains, 45, 112; paraffin-type (polythene) chains, 43, 112; r-distribution, 50; non-Gaussian distribution 102; inverse Langevin distribution, 104; exact distribution, 109; temperature dependence of, 299.
chain entanglements, 74, 168, 228
chemical constitution, 3, 5
cohesive-energy density, 147
cross-linking,
and vulcanization, 11; and modulus, 160 ff.; chemical estimation of, 163 ff.
crystalline polymers, 20
crystallization, 16 ff.
in stretched state, 20 ff.; density change on, 17, 21; internal energy change due to, 39, 283; structural changes on, 19; latent heat effects due to, 39; effects on stress–strain curve, 123.

elasticity,
early theories of, 6; kinetic theory of, 7; of long-chain molecule, 42 ff.; 101 ff., of Gaussian network, 59 ff.; of non-Gaussian network, 113 ff.
entropy,
changes of deformation, 31 ff., 274 ff.; of single chain, 55; of network deformation, 61, 113; of dilution, 129, 133; see also thermodynamics
equivalent random link, 124 ff., 204

Flory-Huggins theory, 136

glass transition, 13 ff.
glassy polymers, 16
Gough–Joule effects, 3

heat of extension, 3, 37, 297

internal energy changes, 32 ff., 274 ff.; in constant-volume deformation, 276 ff; see also thermodynamics

loose end correction, 74 ff., 162

modulus and cross-linking, 160 ff.
Mooney,
equation, 95 ff.; theory, 212 ff.

network theory, 12;
Gaussian, 59 ff., 271 ff.; non-Gaussian, 259, 265
normal stresses in shear, 254; in torsion, 259, 265

Ogden's theory, 233
optical anisotropy, 203 ff;
of chain, 175 ff.; of random link, 205; temperature dependence of, 209; of strained network, 178, 182; of bonds, 203; of monomer units, 204

photoelasticity,
statistical theory, 175 ff.; stress-optical coefficient, 180; effect of swelling, 182; experimental, 189 ff.
polarizability, see optical anisotropy

refractive index and polarizability, 174
resilience, rebound, 15

statistical theory,
of long-chain molecules, 42 ff., 101 ff.; of Gaussian network, 59 ff.; of non-Gaussian network, 113 ff.; value of modulus, 77; of swelling, 136 ff.; deviations from, 95 ff., 227
strain-energy function, 211 ff., 230 ff.;
Mooney, 212; Rivlin's formulation of, 214; Ogden, 233; Valanis-Landel form of, 236
strain invariants, 215, 224, 231, 246 ff.
stress–optical coefficient, 180;
dependence on cross-linking, 195 ff.; for rubber, 196; for gutta-percha,

Printed in the United States
By Bookmasters